THE CANALS OF NORTH WEST ENGLAND

Volume II

THE CANALS OF THE BRITISH ISLES

EDITED BY CHARLES HADFIELD

British Canals. An illustrated history. By Charles Hadfield
The Canals of the East Midlands (including part of London). By Charles Hadfield
The Canals of the North of Ireland. By W. A. McCutcheon
The Canals of North West England. By Charles Hadfield and Gordon Biddle
The Canals of Scotland. By Jean Lindsay
The Canals of the South of Ireland. By V. T. H. and D. R. Delany
The Canals of South and South East England. By Charles Hadfield
The Canals of South Wales and the Border. By Charles Hadfield
The Canals of South West England. By Charles Hadfield
The Canals of the West Midlands. By Charles Hadfield
Waterways to Stratford. By Charles Hadfield and John Norris

in preparation

The Canals of Eastern England. By J. H. Boyes
The Canals of Yorkshire and the North East. By Charles Hadfield

OTHER BOOKS BY CHARLES HADFIELD

Canals of the World
Holiday Cruising on Inland Waterways (with Michael Streat)
The Canal Age
Atmospheric Railways

THE CANALS OF NORTH WEST ENGLAND

by

Charles Hadfield

and Gordon Biddle

WITH PLATES AND MAPS

VOLUME II

DAVID & CHARLES: NEWTON ABBOT

ISBN 0 7153 4992 9

Printed in Great Britain by
Latimer Trend & Company Limited Plymouth
for David & Charles (Publishers) Limited
South Devon House Newton Abbot Devon

CONTENTS

Volume I

VOLUME II

PART THREE—1845–1969

APPENDICES

ILLUSTRATIONS

Volume II

PLATES

TEXT ILLUSTRATIONS AND MAPS

CHAPTER IX

The Manchester, Bolton & Bury Canal

THE initiative for a canal[1] that would give Bolton and Bury water communication with the Irwell, and so with the navigation to Liverpool, seems to have come in the spring of 1790 from a group of men at or near Bolton who commissioned surveys and estimates from Matthew Fletcher. He managed collieries at Clifton and Kearsley, and was to take a prominent part in promoting and running the canal. But this initiative must have been closely linked to the Mersey & Irwell company, for Fletcher had been one of its leaders for many years, Peter Wright their agent sat on the Manchester, Bolton & Bury's first board in 1791, as did a leading shareholder, William Marsden. These, and at least eight other Mersey & Irwell supporters, became original subscribers to the new company, putting in £8,600 between them.

Fletcher's estimates were approved at a meeting at Bolton on 16 September, and by others at Manchester on the 20th and Bury on the 29th. The estimate was £37,437,[2] £31,506 from Manchester to Bury, and £6,931 for the Bolton line from it. The idea looked attractive. A steering committee was appointed, a subscription opened, and Hugh Henshall, veteran engineer of the Trent & Mersey and the Chesterfield canals, and Brindley's brother-in-law, was asked to do another survey and estimate preliminary to a Bill, and particularly to look at likely water supplies, in order to avoid opposition from mill owners depending upon water power.

Henshall reported in October that the canal could be supplied by floodwater reservoirs and rivulet water without interfering with the mills. He estimated for a waterway 5 ft deep and 32 ft wide at surface, with 16 locks, at £42,400. Negotiations were opened with the friendly Mersey & Irwell company, and a hopeful fly was cast over that wily fish the Duke of Bridgewater, that if he 'has not already a power of Locking into the old River so as to communicate with the intended Canal and his Grace shall be inclined to have such powers it shall be provided accordingly'.[3]

The Duke did not rise, but an agreement was made with the Mersey & Irwell which gave them exemption from toll on the canal on all stone needed for bank repairs, and the canal company exemption for their craft unloading on the river between Hunt's Bank and Throstle Nest lock, and favourable tolls below that. £47,700 was then subscribed, the biggest shareholders being the Earl of Derby* and Sir John E. Heathcote[4] with £3,000 each, and Lord Grey de Wilton (later the Earl of Wilton) and Dr James Bent with £2,000.

Charles McNiven did the parliamentary survey, and the Act[5] passed in 1791. A narrow-boat canal was then intended, from the Irwell near the Old Quay or Sugar House to Prestolee, where it would divide to Bolton or Bury. Power was granted to raise £47,700 in £100 shares, and £20,000 more if necessary from shareholders or on mortgage, no one subscriber to hold more than £5,000 worth of shares. No water could be taken from the Irwell, Lever or certain brooks if mill-working would be interfered with. Authorized tolls on coal were 2d (1½d if not passing a lock) as on most other goods except timber and merchandise, for which 3d was the charge. Lime and limestone were 2d, but at half rate should water be running over the weirs. Mineowners were given power to make cuts up to 4,280 yd long, and under this clause Matthew Fletcher made his own canal at Clifton (see p. 261).

At the first meeting Charles Roberts of London was appointed engineer, seemingly under Fletcher as superintendent in general charge of construction, and with Hugh Henshall as consultant. In September 1793, Roberts was found 'not to have acted in the Execution of his Duty with proper discretion and Oeconomy',[6] and was dismissed, being replaced by Matthew Fletcher's nephew John Nightingale. The canal was begun, not at its junction with the Irwell, but 3 furlongs away at Oldfield Road, below the first of the six proposed connecting locks, and cut thence upwards and along the branches to Bolton and Bury.

The company were much distracted in their early years by possibilities of extension that opened before them, towards the

* Four miles of the canal ran through his land. His agent, James Wareing, took an active part in the company's early years.

Calder & Hebble and Halifax and towards the Leeds & Liverpool at Red Moss near Wigan.

Brindley, in his original survey of 1766 for a Rochdale canal (see Chapter X) to Sowerby Bridge, had suggested alternative lines from Manchester via Heywood or via Bury. This latter idea had been picked up in 1770 by the Liverpool Canal promoters, who saw a possible extension from their proposed line towards Bolton, Bury, Rochdale and Halifax.[7] The Rochdale idea was revived early in 1791, but the promoters had not decided upon their line when the Manchester, Bolton & Bury got their Act. It was therefore an obvious thought that the latter should either attract the Rochdale company to join their line at Bury instead of seeking an independent entrance to Manchester, or extend their own canal upwards past Bury to the summit by Littleborough or even to Sowerby Bridge and so eliminate the Rochdale scheme.

In September 1791 notice was given of a canal from Bury to Rochdale, Chelburn and Dean on the summit.[8] This may have caused the Rochdale committee in October to approach the Manchester, Bolton & Bury. They seem to have been put off by their reception, for when Rennie put a line via Bury to shareholders in December, he was told that the committee ruled it out. A group of Manchester, Bolton & Bury supporters led by Fletcher and the Earl of Derby then promoted the Bury & Sladen Canal* early in 1792 to join the Rochdale near Littleborough. It was surveyed by William Bennet[9] and estimated at £40,376.

When the Rochdale's Bill of 1792 failed, the Bury & Sladen saw their chance, and asked the Rochdale's supporters to join them instead. These refused, and decided to promote a Bill for a canal from Manchester only to Chelburn on the summit not far from Sladen. The Bury & Sladen promptly gave notice of an extension to Sowerby Bridge, as then did the Rochdale in self-defence.

The position was dangerous for the Rochdale, who naturally opposed the Bury & Sladen extension desperately. Longbotham had surveyed the latter, which was not to take the Rochdale's proposed line via Todmorden and Hebden Bridge, but was to cut through a long tunnel to the Calder near Halifax, at an estimated cost of £190,291. The Rochdale supporters pointed out that this involved 8,428 yd of tunnelling (at the time the Rochdale was proposing a 3,000 yd summit tunnel on its own line, for which

* Sladen is near Littleborough.

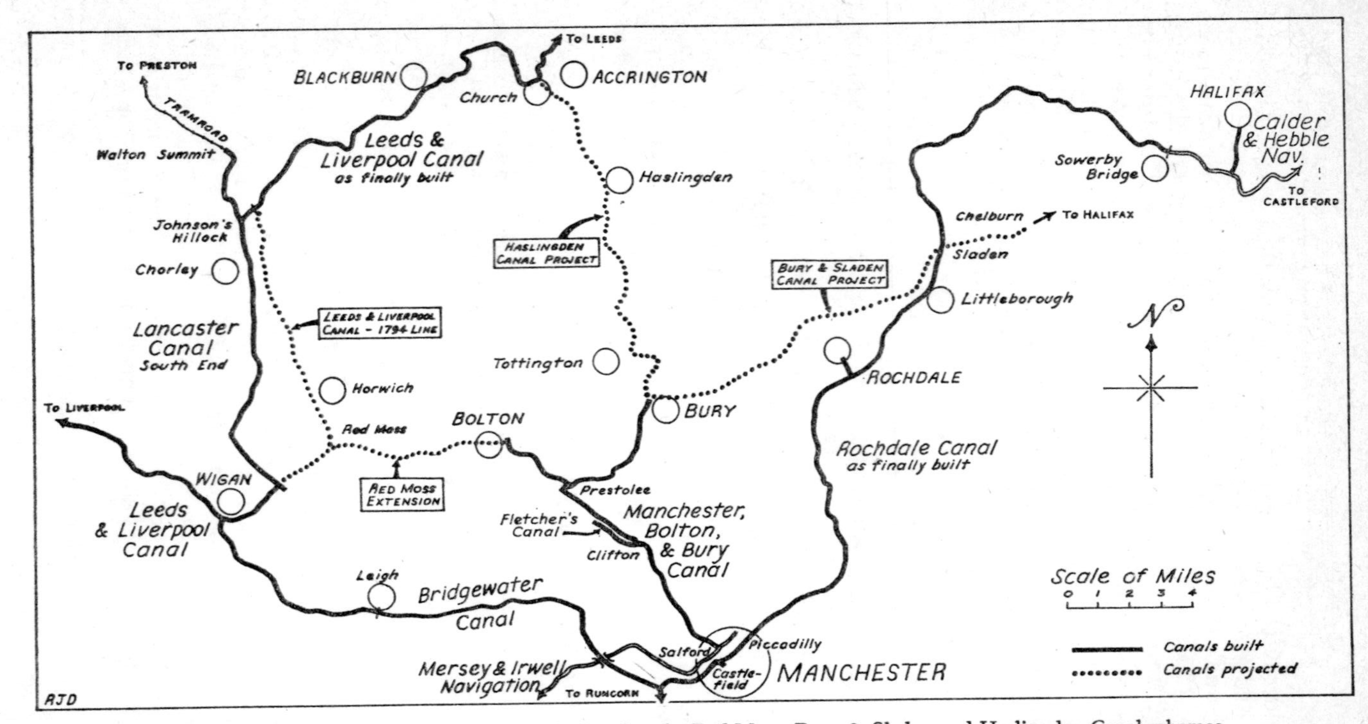

16. The Manchester, Bolton & Bury Canal, showing the Red Moss, Bury & Sladen and Haslingden Canal schemes

locks were substituted later), that it would take the waters of the Roch and so interfere with industry, and that far more landowners had dissented to it than to their own line.[10]

Meanwhile, the parliamentary line of the Leeds & Liverpool Canal as it approached Wigan seemed unlikely to be followed, and one survey for a possible alternative took it from Blackburn by way of Chorley and Red Moss (between Horwich and Blackrod) to Wigan. The Manchester, Bolton & Bury saw that, if their Bolton line were to be extended for 4½ miles to Red Moss, they would gain all the traffic arising between the Leeds & Liverpool and Manchester (see Chapter VII). A separate group promoted this extension, estimated at some £18,000, and some land was bought. In 1792 they amalgamated with the Bury & Sladen supporters to introduce a joint Bill in 1793. Fortunately for the Rochdale, it failed. Negotiations with the Leeds & Liverpool company continued, and another joint Bill was put forward in 1794. On the Sladen extension it was a neck-and-neck affair with the Rochdale, for the two were originally down for second reading on the same day. But there the race ended, for the extension Bill failed by 36 votes to 13,[11] and the Rochdale's went on to success. The Sladen extension was now dropped, but the promoters held to the Red Moss line. These latter must have given the Manchester, Bolton & Bury an impression of hostility, for this company tried to take over the scheme later in 1794, but were refused.

The Manchester, Bolton & Bury tried to find out whether the Leeds & Liverpool were firmly for the Red Moss line and their Bolton link, because if so they would have to consider whether to build wide instead of narrow locks. A delegation was sent, which interviewed Joseph Priestley and two others, and came back to say that these, while not binding the Leeds & Liverpool, considered as individuals that their company intended to build the deviated Blackburn–Red Moss–Wigan line 'with all Spirit and Dispatch' if they could get the money, and did not think it had considered any other junction with Manchester than by the Manchester, Bolton & Bury. The Leeds & Liverpool three added that were the Manchester, Bolton & Bury to become a broad canal and extend to Red Moss, this would be an added inducement to the Leeds & Liverpool.

At this time, as we have seen, the Leeds & Liverpool had been built from the Aire & Calder at Leeds to Gargrave. Cutting thence towards Lancashire had re-started in 1790, but it was still a long way from Red Moss. One cannot therefore altogether blame

Priestley and his companions for being uncertain about a route that could not be built for many years yet.

On the strength of this interview, however, and of the existing agreement, a shareholders' meeting decided to convert the part of the Manchester, Bolton & Bury that had already been built to take broad instead of narrow* boats, and to build the rest to the lock and bridge-hole widths of the Leeds & Liverpool at an extra cost of £11,000. In fact, it cost much more: over £20,000, according to a company statement of 1809.[12]

Meanwhile, work had been going on, and by mid-June 1794 cutting was complete from near Oldfield Road past the top lock there and for about 5 miles on the level to Rhodes lock. It was this section, and the stretch onwards for about a mile to Giant's Seat, that had to be converted, together with some outlying structures such as the aqueduct near Bolton. Soon afterwards, £100 having been called on each share, an attempt was made to raise money by mortgage. This failing, extra calls were made on the shares. By the beginning of 1796, with £67,196 raised out of the authorized total of £67,700, Benjamin Outram went over the works, and reported that £26,924 would be needed to finish them on the enlarged scale, including the Irwell junction. The shareholders at first proposed to seek an Act to authorize more capital, but then, fearing trouble from millowners over water supplies, some decided to raise another £7,000 themselves by voluntary additional calls of £22 10s to get the line open upwards from Oldfield Road, even though only to 4 ft depth, 'so as to render the Canal useful to the public'.[13]

The first few pounds of tolls had been taken in 1795. By mid-1796 the section from Oldfield Road to Rhodes lock was in use by barges carrying 40 tons and narrow craft with 20. The line through to Bolton was navigable about the beginning of October, for on the 13th a packet boat started its daily return trip from Bolton to Manchester. The Bury line had first been used three weeks before, on 24 September,[14] though in June 1797 it was still only navigable by narrow boats. However, a good deal of work was still needed over the whole line, while the Irwell connection

* Though the Ashton-under-Lyne Canal had been authorized in 1792, there was no connection between it and the Mersey & Irwell. The Manchester, Bolton & Bury was therefore designed to take somewhat shorter narrow boats than the standard 70-footer. After widening, the locks took 68 ft long barges or pairs of short narrow boats 68 ft long, but when, later, 70-footers began to use the canal extensively, it seems that only one, lying slightly askew, could be worked through the locks at a time.

had not been begun. Expenditure had been £80,906 to 23 June 1797, and a call of £1 18s 6¾d (£1·93) was made to clear debts. Shareholders who had paid all the calls, including the voluntary ones, had subscribed £164 8s 6¾d (£164·43) per £100 share, others £140. The position was awkward. The sensible thing was to go to Parliament for an Act to clear up the existing differences between what shareholders had subscribed, and give powers to raise enough new money to finish the canal and put it into good trading order. But, now the line was open, millowners were watching for any infringement of the water clauses of the 1791 Act, were threatening to introduce their own Bill[15] and would certainly cause trouble in Parliament. Therefore another voluntary call of £20 was made in 1802 to finance a reservoir at Radcliffe and the Irwell extension, and no Bill was introduced until 1805. Instead, profits were ploughed back, and landowners asked to provide wharves and warehouses.

The company still had hopes of a connection with the Leeds & Liverpool at Red Moss, and in 1797 gave the extension subscribers the choice of starting work within one year and finishing it in five, or handing over their land agreements and letting the Manchester, Bolton & Bury do it. Whereupon the scheme was relinquished to the company. In 1799 the Manchester, Bolton & Bury asked for a meeting with the Leeds & Liverpool, saying that they were 'preparing to make Locks at Bolton for the Red Moss line',[16] and in 1800 included it in a Bill. Then, the Leeds & Liverpool's line still far from completion, their attention switched to an alternative way of ending their isolation.

Not far away from their terminus, the section of Rochdale Canal between the Bridgewater at Castlefield and Piccadilly was nearly ready. The Ashton-under-Lyne Canal (see Chapter XI) was also approaching Piccadilly, and was itself being extended by way of the Peak Forest and Huddersfield canals (see Chapter XII) towards the limestone quarries of the Peak and the trade of Yorkshire. The Rochdale's own line, finished from the Calder & Hebble at Sowerby Bridge to Rochdale at the end of 1798, was also creeping towards Piccadilly. But the Mersey & Irwell was not joined to any of these,* and the Manchester, Bolton & Bury's committee therefore began to wonder whether, instead of joining the Irwell, they would do better to cross it on an aqueduct to meet

* The Mersey & Irwell was joined to the Rochdale Canal by way of the Manchester & Salford Junction Canal on 28 October 1839, and to the Bridgewater when the Hulme locks branch of the latter was opened on 20 September 1838.

the Rochdale between Castlefield and Piccadilly. They decided to sound the Rochdale company and the Duke, and in 1799 gave notice of a Bill. The Mersey & Irwell reacted so strongly, however, that they dropped the idea, and turned to settling the water problem, for any junction, whether with the Irwell or other canals, would lead to heavy water losses as traffic increased. By July 1801 they had settled on a reservoir at Radcliffe, to be supplied with floodwater from the Irwell, and on using surplus mine water where possible.

When in November 1801 they again took up the question of a junction, they no longer talked of by-passing the river, but, instead, of a tramroad from Oldfield Road to the Irwell, and sent a deputation to the Old Quay company to suggest a 'navigable Tunnel from the Old River Navigation towards the Rochdale Canal', a foreshadowing of the later Manchester & Salford Junction Canal. Then, in June 1802, they started to buy land to make the Irwell junction as originally planned. But even then it took some time to obtain all the necessary land, and the connection was not completed until December 1808.[17]

The total cost of building the canal seems to have been about £115,500. From the Irwell the line rose by six locks (four being staircase pairs) to the basin at Oldfield Road, Salford, where two arms were later built, a short one almost to Hulme Street, and a longer one eastwards and parallel to it and to Oldfield Road, as far as Windsor Crescent (then Broken Bank). It then ran along the Irwell valley past Fletcher's Canal to the stone-built three-arched Clifton aqueduct over it. Then past five other locks and the four-arched stone Prestolee aqueduct, also over the Irwell, to the foot of the Prestolee flight of six locks in two staircases of three. Above them the canal divided, one branch running to Bolton across the three-arched brick aqueduct over the Tonge, the other to Bury, both on the level. The length from the Irwell to the top of Prestolee locks was 8 miles: then 3 miles to Bolton and 4¾ to Bury. There were originally two tunnels at Bury, one of 66 yd, really a road bridge, and another of 141 yd that would only take narrow boats, beyond Bury basin on a short extension called the Coal Branch that led to other wharves. The reservoir was about a mile from Bury.

In 1805 the company at last obtained an Act[18] to settle their finances. Under it they were empowered to raise £80,000, in shares or on mortgage. Some £52,470 was raised in share capital, and £16,000 on mortgage, under this Act, part of the money being used to repay shareholders who had paid excess calls, so restoring the nominal value of all shares to £100. In June 1812 the first dividend was declared, of £4 a share. The rate rose to £5 in 1813 and £6 in 1814, and then moved as follows:

Year	*Dividend per cent*	*Year*	*Dividend per cent*
1815	6	1823	5
1816	6	1824	6
1817	5	1825	6
1818	5	1826	6
1819	4	1827	6
1820	4	1828	6
1821	5	1829	6
1822	5	1830	6

In 1808, a few months before the Irwell connection was opened, the company were considering whether to narrow the Irwell locks to save water. They did not, and at the end of 1811 agreed with the Mersey & Irwell company that the latter would carry over their line between Bolton, Bury and Liverpool, which they did for more than twenty years. A number of other carrying firms also worked on the canal. Wide boats conveyed much of the traffic, though short narrow boats, 68 ft long, predominated, some holding nine 35 cwt to 2 ton coal containers.

In 1809, with the Leeds & Liverpool nearly completed downwards to Blackburn, the Manchester, Bolton & Bury heard of the former's Bill for a branch from Wigan to Leigh to join the Bridgewater trustees' Leigh branch, and knew they had been worsted. Indeed, a clause in the Bill restrained the Leeds & Liverpool from making any other connection with Manchester without the trustees' consent. The Manchester, Bolton & Bury regarded the Bill as incompatible with their 1794 agreement with the Leeds & Liverpool, and succeeded in defeating it in the Lords. In 1811, however, the Leeds & Liverpool decided to vary their line to join the Lancaster Canal at Johnson's Hillock and use it thence towards Wigan. This was done, the Leeds & Liverpool was completed in 1816, and in 1819 put forward another Bill, this time successfully, for the Leigh connection.[19] After angrily consulting counsel upon possible legal action for breach of contract, the Manchester, Bolton & Bury gave up, and sold their land on the Red Moss route.

In 1824, however, the Manchester, Bolton & Bury agreed to Ralph Boardman's proposal for a railway from the Leeds &

Liverpool at Leigh to their canal at Bolton. Ralph Boardman was town clerk of Bolton and clerk to the proposed railway company, which had been formed to seek a transport outlet for Bolton goods and passengers to Liverpool. The Manchester, Bolton & Bury agreed to it on condition that it joined their canal as well as the Leeds & Liverpool, presumably calculating that traffic from Bolton lost to them would be balanced by gains coming from the Leeds & Liverpool. Their agent, Robert Darbishire, was a subscriber, as were a few of the bigger shareholders, notably Peter and Richard Ainsworth, Peter Rothwell and John Pilkington.

The Bolton & Leigh Railway's Act of 1825 authorized the line 'from or near the Manchester, Bolton and Bury Canal, in the parish of Bolton-le-Moors, to or near the Leeds and Liverpool Canal, in the parish of Leigh'. Liverpool being the aim, it seems that the junction at Leigh was regarded as a temporary expedient, and a junction with the Liverpool & Manchester Railway, then being promoted, the final aim. The railway was partially opened from Bolton (though the connection to the canal was never made) in 1828, and through to the Leeds & Liverpool Canal in 1829, but by early 1831 trains were instead crossing the canal and running by way of the Kenyon & Leigh Junction Railway to the Liverpool & Manchester[20] (see p. 168).

So, in the end, the Manchester, Bolton & Bury got no benefit from the Leeds & Liverpool connection. But by then they were themselves promoting their own railway from Bolton to Manchester.

Their line completed in 1808, the company settled down to modest and uneventful prosperity under the management of Robert Darbishire as engineer, agent and assistant clerk. The main traffic was coal, down to Manchester and up to Bolton and Bury, from collieries at Clifton and elsewhere,[21] building and road stone, flagstones and limestone, and groceries and sundries traffic upwards from the Mersey & Irwell. Darbishire stayed until 1830. Hardly any traffic figures are known, except that the years 1835–7 show some 120,000 tons of coal a year passing downwards to Oldfield Road.

Passenger-carrying was a feature of this canal. The packet-boat that started to run in October 1796 was operated by the company. It then left Bolton for Manchester at 06.00 (summer) and 07.00 (winter), returning at 17.00 (16.00). The fare was 1s 6d (7½p) single, 2s 6d (12½p) return in the state cabin, or 1s (1s 6d) (5p and 7½p) in the after cabin. The boats also carried parcels, small ones at 6d (2½p) each, large at 6d (2½p) per 60 lb. Passengers changed at Prestolee from one boat to another, to save having to wait while six locks were worked, by walking under a covered way.

The service was successful, and further craft were built. In 1804 the company decided no longer to work the craft themselves, but to let them by tender. James Barnes took them for two years at £1,410 p.a.; thereafter the boats were sometimes worked by lessees, usually Barnes, sometimes by the company if they did not think the outside bids were high enough. Rents did not vary much from the £1,400 mark. By autumn 1804 Barnes was running two services a day each way.

In 1810 an additional service began between Bolton and Bury along the summit level. Fares were raised in 1814 to 2s (10p) and 1s 4d (6½p), but lowered again in 1822 to 1s 4d (6½p) and 1s (5p), by 1830 to 1s 3d (6p) and 10d, and by 1832 to 1s (5p) and 9d. In 1815, when boats were leaving Oldfield Road for Bolton at 07.00, 17.00 (except Tuesdays and Saturdays) and 18.00 (summer Tuesdays), the company stated that 'the time is *strictly* observed, being regulated at Manchester by St Ann's Church Clock and at Bolton by the Bolton Church Clock'.[22]

In June 1818 there was a nasty accident, with several deaths when the Bolton and Bury boat was upset by a party of drunks.[23] About 1825 a young woman described a journey she was accustomed to make on these packet-boats:

> 'our trunks were carried to the boat by the two maidservants; and after a grand flourish on the horn and a loud "Gee"! from the captain we set off; two horses pulled at the rope, and we went on our way at a speed of three miles an hour. The boat was covered and had seats and a table inside; but if the weather was fine it was pleasant to sit in the forepart outside. . . . A young woman in curlpapers and a coral necklace came round to know if any of the passengers would take breakfast, which she could prepare in a little cabin that separated the best part of the boat from the other. We never took it because it cost a shilling more, but we sometimes expended twopence in currant cakes or peppermint drops that were carried round in a basket'.[24]

In 1832 the company built and tried a light iron swift boat, the *Lancashire Witch*, built by Fairbairn's, using four horses, but could not get the expected speed of about 10 m.p.h. So they decided to work it more slowly, though still cutting the time from Bolton to Oldfield Road from 3 to 2½ hours with less work for the horses. In the year ending 31 October 1834 the Bolton to Manchester packets earned a net £1,177 and the Bolton to Bury boat another £75, these receipts being over 10 per cent of the company's income of £11,360 for 1834. In the twelve months from July 1833 to June 1834 these passengers were carried:

Bolton to Manchester	21,060
Manchester to Bolton	21,212
Intermediately	20,818
Race traffic (May)	1,324
	64,414

The best months were May, with 8,034 passengers including racegoers, and July with 6,765; the worst, February with 3,748 and November with 4,145.

In 1836 there was a boat from Bury to Bolton only on Sunday (summer) and Monday mornings at 08.30, returning at 17.30.[25] From Bolton to Manchester boats left on Tuesdays and Saturdays at 06.00 and 17.00, on other days at 07.00. In May 1838 the railway was opened from Bolton to Salford, and the boats ceased to run and were sold.

In July 1825 a delegation went to see R. H. Bradshaw about a possible junction with the Bridgewater Canal, and must have got some encouragement, for they then had the levels surveyed. Nothing, however, happened for the time being. Instead, the company watched the building of the Liverpool & Manchester Railway with anticipation rather than fear, for in March 1829 they had in mind building a branch railway from their canal at Oldfield Road to join it. Their ideas moved quickly on to building such a line right back to Bolton. By late July 1830 £66,700 had already been subscribed[26] and in September they decided to employ 'Mr Stevenson or some other professional Engineer'[27] to survey the whole line of canal, and report how practicable and expensive it would be to convert it to a railway with the minimum of deviation. They got Alexander Nimmo to do the job. He reported a fortnight later that conversion was possible, and gave an estimate, 'so far as he expressed himself capable of judging from his present cursory view of the Canal'.[28] The shareholders then decided to seek a Bill for a railway from Bolton to Manchester.

However, a deputation from a proposed company to build a railway from Liverpool that might also be extended to Leeds,[29] headed by Sir John Tobin, attended the same shareholders' meeting, and in the course of it offered to buy the canal for £200 per share cash or £300 in railway shares, mortgages to be taken over by the railway company, the offer only to apply if both companies got their Act. This offer the shareholders accepted in principle.

The company's Act[30] was eventually passed on 23 August 1831, their name now being changed to the Company of Proprietors of the Manchester, Bolton & Bury Canal Navigation & Railway, with power to make a line from Manchester to Bolton and Bury 'upon or near the Line of the . . . Canal', and a branch from Clifton aqueduct through Clifton, Kearsley, Farnworth and Great Lever. This line, as far as the Clifton and Kearsley collieries, was to be opened, together with a single-track railway from Bolton to the Irwell, however temporary, before the canal closed, to provide continuity of transport. Otherwise, they were authorized six months after the Act to close parts of the canal needed for the railway, except between Bolton and Bury, and from the Irwell to the interchange basin with the railway at Oldfield Road. The capital was £204,000. Because the canal line was to be followed closely, there was to be a cable-worked incline at Prestolee. However, there had been considerable opposition to the Bill from coal dealers who saw themselves losing access to the canal for their supplies without having railway branches provided for them; these were only placated by the company's agreement to seek an amending Bill, to keep the canal open and build the railway alongside it.

The Manchester, Bolton & Bury company had got their Act, but Tobin's group had not. Between 23 August and 22 October 1831 (when possession was given), however, Sir John Tobin and his Liverpool supporters bought almost all the 477 canal shares, paying £200 cash for 279 of them, and rather less for others whose owners proposed to subscribe to the railway that was to parallel the canal. It seems that in the end the new group paid £86,231 in cash, and took over mortgages for £1,976.

Of the 477 shares, Tobin and his group bought all except eleven, eight of which were kept by old shareholders,* and three belonged to the company itself, having been forfeited in the past. The 477 were then multiplied by thirteen to a total of 6,201 new shares, 104 being issued to replace the eight, and 39 the three; 5,295 were issued to subscribers, the balance being held by trustees until they were later issued pro rata.

A new Act[31] was passed in 1832 to provide a railway route separate from the canal which had now to be maintained for ever,

* These were bought in 1834.

though it could be somewhat straightened in places. Nimmo having died, Jesse Hartley, who was surveyor and engineer of Liverpool docks, was appointed engineer, and in 1833 railway construction began. The shareholders were told that

> 'by progressive and judicious repairs and alterations in the Canal, attended with care and economy in its management, the Proprietors may reasonably calculate on an improving revenue independent of the advantages to arise from the projected Railway'.[32]

A number of wharf and other improvements were in fact made, the Mersey & Irwell pressed for toll reductions, some in the interests of the Liverpool trade, and the Peak Forest Canal granted drawbacks on lime from its canal carried to the summit level of the Manchester, Bolton & Bury.[33] Dividends were paid from canal profits at 5 per cent.

In 1836 the company were still building their railway to Bolton, but on a new line from Clifton along the hillside above the left bank of the Irwell to Bolton, divorced from the canal. They had agreed with the Liverpool & Manchester Railway for a junction at Salford, and with the Bolton & Leigh for one at Bolton, so that they could have through railway communication to Manchester for their coal, and via the Kenyon & Leigh Junction Railway to the Liverpool & Manchester's line to Liverpool.† They would now be much less dependent on their canal's connection with the Mersey & Irwell. All the same, they welcomed the promotion of the Manchester & Salford Junction Canal to link the Mersey & Irwell to the Rochdale Canal, and so to the Ashton, Peak Forest and Huddersfield lines; 'from this source', said the committee, they 'confidently expect a large amount of new Traffic on the Canal'.[34]

The railway to Bolton was opened, with not too many interruptions to canal traffic. After a good deal of construction material had been carried, services began for passengers on 29 May 1838 and for goods early in 1839. No action was taken about the branch to Bury. In January 1839, in anticipation of the opening of the Manchester & Salford Junction Canal, the Ashton and Peak Forest companies sent delegates to the Manchester, Bolton & Bury 'to urge the necessity of lengthening such of the Locks upon that Canal as are not of sufficient length to admit of the Boats from the Ashton and Peak Forest Canals to pass the same',[35] but this was not done. Some changes had been made in the canal line to

† The L.M.R. junction was made in 1844, the Bolton & Leigh not at all.

accommodate the railway. At Oldfield Road, for instance, a lock was moved, 150 yd of canal lowered, and two short tunnels of 34 and 49 yd built.[36]

It seems to have been planned that the canal should keep its coal and limestone traffic, for which cheap maximum tolls were authorized in July 1840, and that the railway should carry passengers and attract merchandise traffic from the roads. In July 1839, however, the chairman sadly reported that it was now six months since the railway had opened for merchandise carrying, but little had yet come from the road. Indeed, there was a time in 1843 when the company were discussing the improvement of their canal with the Mersey & Irwell.[37]

In December 1844 the Manchester, Bolton & Bury agreed to amalgamate with the Manchester & Leeds Railway; this was approved by the shareholders in May 1845 and ratified by an Act[38] of 1846 which authorized new and lower maximum canal tolls of ¾d on coal, and 1d on almost everything else except timber and merchandise. A year later it had become part of the Lancashire & Yorkshire Railway. (*To continue the history of the Manchester, Bolton & Bury Canal, turn to p.* 439.)

The Haslingden Canal project

Following the rejection of the Leeds & Liverpool's first deviation Bill in 1793, which, it will be remembered, had been supported by the Manchester, Bolton & Bury in return for an undertaking to make a junction with their proposed branch to Red Moss, another scheme to join the two canals was mooted, with indirect Manchester, Bolton & Bury support. This time the proposal was from Bury to the Leeds & Liverpool at Church, near Accrington, via Haslingden, about 13 miles. Fletcher and McNiven, engineers to the respective canal companies, had made a preliminary survey and pronounced the scheme practicable, despite a climb over the intervening watershed of some 400 ft, and detailed work was then done by William Bennet. The Leeds & Liverpool saw the projected canal primarily as a means of water supply and the agreement of the previous year with the Manchester, Bolton & Bury for water from the Red Moss branch was to be modified accordingly. The 'Haslingden branch' was seen as an additional connection with the Manchester, Bolton & Bury rather than an alternative to the Red Moss branch, and the Leeds & Liverpool's committee resolved that the public should be informed that two junctions were now proposed. Parliamentary lobbying was put in hand, and the two companies agreed to 'obtain' a petition from the Duke of Buccleugh's tenants at Haslingden.[39]

Shortly afterwards the Haslingden Canal promoters asked the Leeds & Liverpool, should their second deviation Bill fail, to extend on their original parliamentary line to near Hyndburn Bridge, between Padiham and Whalley, and they would apply to do the same; this was agreed.[40] No reference to the project, however, appears in the Manchester, Bolton & Bury's records.

Haslingden itself, a market town perched on the top of a hill, was becoming a manufacturing place and a centre for stone quarrying, and a certain amount of traffic was in prospect. 'Some bulky goods, as oil for the woollen industry, are brought from London by the Selby Navigation into Yorkshire, and hither (to Haslingden) by land carriage.'[41] An Act[42] was obtained in April 1794 for a canal from Bury bridge through Elton, Lower Tottington, Higher Tottington, Haslingden and Accrington to Church Kirk, and although a junction was authorized with the Manchester, Bolton & Bury the Act makes no reference to one with the Leeds & Liverpool, despite the clear intention expressed in the latter's minute books. There were to be no locks without the consent of three-quarters of the owners of 'Mills, Factories, and other water Works . . . on the River Irwell between Ratcliffe (sic) Bridge and Heywood Bridge, and upon the Haslingden Stream, and upon the Stream or Streams running from, or from near Carter-Place, to Church'. Otherwise changes in level were to be achieved by 'Rollers, Racks, Inclined Planes, or any other Works, Engines, or Machinery, as a substitute for Locks'. Severe restrictions on water supplies were also incorporated. The authorized capital was £47,600 in £100 shares, with provision for raising an additional £40,000 by new shares or mortgage of the tolls.[43]

In July 1794 the committee asked millowners on the Irwell and other rivers affected to discuss a scheme for conveying boats from one level to another without the use of locks, 'by which Means a very small Quantity of Water will be wanted, and may be supplied without injuring the Mills in the smallest Degree'.[44] Thereafter, the scheme faded away, and it seems likely that the Manchester, Bolton & Bury pressed on with the Red Moss connection with the Leeds & Liverpool in order to discourage it.[45] The concern was wound up in 1797[46] along with the Red Moss scheme.

The restrictions on building locks, which implied inclined

planes or lifts, also implied a tub-boat canal.[47] Even then, the necessarily short summit and heavily restricted water supply would have made it difficult to work. To what extent the Haslingden scheme was genuine, or just a move in the canal politics of the time, is difficult to judge. But there still remains a sharp, awkward right-angled bend in the Leeds & Liverpool at Church warehouse, that tantalizingly suggests it was intended as a point of junction with the Haslingden Canal.

Fletcher's Canal

In the 1750s, John Edensor Heathcote, the owner of the Wet Earth colliery at Clifton, and son-in-law of Sir Nigel Gresley of Apedale, called in James Brindley to advise him on how best to prevent flooding of the workings. Brindley devised an interesting answer which involved using the water of the River Irwell to provide power to drive a paddle wheel to lift water from the colliery, and included in his works a ½-mile surface channel between the river and the colliery.[48] By 1760 Matthew Fletcher, formerly Heathcote's mining engineer, who was working the mine, had sunk another pit, the Botany Bay, a mile east of Wet Earth. At the same time he extended Brindley's channel for this distance and installed another paddle wheel, in this case geared to wind coal.

Probably about 1790–1 Matthew Fletcher, who was also a committeeman of the Manchester, Bolton & Bury Canal, deepened and widened these channels to convert them to a canal to get his coal away, while still keeping their earlier purpose. The part of the channel between Brindley's wheel and Wet Earth was, however, only used as a lay-by. Although the Manchester, Bolton & Bury was opened to Salford in October 1796, in July 1799 there was still a 30 yd gap between it and Fletcher's Canal. There seem to have been two troubles: first, because of the strict conditions under which millowners on the Irwell had agreed to the Manchester, Bolton & Bury taking water higher up, it was essential that no water should be lost to Fletcher's Canal. Secondly, Fletcher seems to have built a lock in the gap at the end of his canal ready for the junction, which had too great a fall for the final water level of the Manchester, Bolton & Bury. Outram, advising the canal company, apparently recommended a second lock at the main canal end of the gap, with a small rise. This suggestion, though at first approved by the canal committee and the millowners, does not seem to have been acted upon. Instead, the original lock seems to have been widened and extended across the gap to form a single large lock, 90 ft long and 21 ft wide, so taking three narrow boats at once, with a fall of 1 ft 8 in. towards the Manchester, Bolton & Bury, which therefore always gained water. Fletcher's Canal was 1½ miles long, and joined the main canal line near Clifton aqueduct, having run parallel to it on the opposite bank of the Irwell.

It seems certain that Fletcher started to build underground canals after the Worsley example both at Wet Earth and Botany Bay. At Wet Earth a plan of the Doe mine workings there, dated 1860, shows an 'old boat level' about 1,000 yd long running to the estate boundary, its probable limit. Its entrance was also that of Brindley's wheelrace. At Botany Bay, where the entrance was later obliterated by the Lancashire & Yorkshire Railway, the line seems to have run for about 2 miles to the Spindle Point colliery.[49] Fletcher himself died in August 1808, aged 77. In the 1810s the canal's water ceased to wind coal at Botany Bay, but continued to drain the mine.

When the Manchester, Bolton & Bury company were empowered in 1831 to convert their canal to a railway, the Act provided for a rail branch to Clifton and Kearsley collieries, prohibited the company from using Fletcher's Canal for the line, and gave him free carriage on the railway branch because his canal would be useless. However, the 1832 Act which continued the canal repealed Fletcher's exemption.[50] (*To continue the history of Fletcher's Canal, turn to p.* 441.)

CHAPTER X

The Rochdale Canal

ON 2 July 1766 the Bradford meeting was held that launched the Leeds & Liverpool Canal (see Chapter VII) from the Aire & Calder Navigation at Leeds (which in turn gave access to Hull) by a long and winding route that missed out Manchester altogether. Richard Townley of Belfield near Rochdale was probably the moving spirit[1] in calling another meeting at the Union Flag inn, Rochdale, six weeks later, on 19 August, to promote a rival canal 'to join the east and west seas' from the Calder & Hebble Navigation[2] open from Wakefield to near Salterhebble, and building towards Sowerby Bridge, to either the Mersey & Irwell Navigation or the Bridgewater Canal, then being built towards Runcorn. It was a shorter line if long-distance traffic were the main motive, but served fewer intermediate towns and industries than its competitor. The gathering supported the plan, and afterwards 47 subscribers put up money for a preliminary survey, among them names like Royds, Smith, Wood and Hamer that were to recur in 1791.[3] James Brindley then did two surveys, one similar to the later canal, to cost £79,180, and one via Bury at £102,625.[4] In December there was also talk of a branch from Todmorden to 'the limestone rocks of Salterforth and the southerly parts of Craven', an area whose limestone quarries were also interesting the Leeds & Liverpool promoters.[5] Perhaps the severe flood damage to the Calder & Hebble in October 1767 prevented further action. Certainly the project slept for twenty years.

The building of the Leeds & Liverpool had begun at both ends, until by 1790 it had reached Gargrave in Yorkshire and Wigan on the Lancashire side. Meanwhile, merchandise was being carried over the Pennines from Lancashire by road waggon, one route being to Sowerby Bridge. An advertisement of 1788 shows that it was taken thence by way of the Calder & Hebble and the Selby Canal[6] (which had been opened in 1778) to Selby, to be transhipped to river sloops for carriage to Hull. Textile raw materials were the principal back-carriage to Lancashire. Through freight charges, including tolls, were £2 5s 2d (£2·26) a ton from Manchester to Hull, and £2 4s 8d (£2·23½) from Hull to Manchester.[7]

In 1790, when canal-building was in the British air, a meeting at Hebden Bridge considered extending the Calder & Hebble there. Rochdale men attended, however, and suggested a through canal to Manchester. Another meeting was called at Rochdale on 17 February 1791

> 'to take into Consideration the desirable advantage, likely to arise . . . by extending the two Navigable Canals from Manchester and Sowerby Bridge Wharfs, so as to make a Junct. thereof near this place'.[8]

It adjourned to April, and meanwhile anyone who had copies of Brindley's surveys was asked to look them out. The April meeting approved a line by way of Todmorden, and appointed a committee which included George Travis, archdeacon of Chester, chairman of the meeting and thenceforward to be a leader in carrying out the project. It was hoped that William Jessop would do the survey, or, failing him, Robert Whitworth; neither could spare the time, and in June John Rennie, who had then no canal-building experience, was chosen, with William Crosley senior of Brighouse, who had already done preliminary surveys, to help him. The subscription, set at £200,000, had all been raised by early June, most of it coming from within or near Rochdale. There was also considerable Manchester support, and a little from neighbouring towns. Only a few names had Yorkshire addresses.

In August Rennie was told to survey branches into Rochdale, to Oldham, and from Todmorden to limeworks in Craven on the proposed line of the Leeds & Liverpool company. The Manchester termination was unsettled, and by early September the committee had decided to ask the Duke of Bridgewater and the Leeds & Liverpool for leave to join their canals. In that month also the Ashton-under-Lyne Canal was promoted, which affected Rochdale plans, because it was to run towards Oldham and would therefore compete with the proposed Oldham branch.

The Duke refused a junction, and the snubbed promoters went sadly back to him to ask whether their canal might 'be extended to a point so near his Navigation, that the Goods Etc. transported on those Canals might be unloaded from the Vessels on the One, into those on the other by means of a Crane'.[9] They also decided to talk to the Mersey & Irwell company about joining them instead. At this point the Manchester, Bolton & Bury company (see Chapter IX) came on the scene. Their Act had been passed in this same year of 1791 for a canal from the Mersey & Irwell Navigation in Salford to the two towns, and a group of their shareholders were now pressing for extensions, one to Sladen on the Rochdale's summit level near Littleborough, the other to the Leeds & Liverpool's future line at Red Moss near Wigan.

The Bury & Sladen scheme (see p. 247), later extended to include a line right through to Sowerby Bridge, was a direct rival to the Leeds & Liverpool and the Rochdale and, because it was supported by the Mersey & Irwell company as a means of extending their own trade, it made the Duke more inclined towards the Rochdale project, for he certainly did not want his own canal line to Runcorn for Liverpool trade to be by-passed. He now said that he would make up his mind when he had seen the Rochdale promoters' draft bill. Rennie advised them to get as close to the Bridgewater Canal as they could, and 'I doubt not but in time you will force your way into the Duke's Canal in spite of his opposition'.[10]

After Rennie's survey, the Rochdale suggested to the Leeds & Liverpool a branch from Todmorden by Holme Chapel and Worsthorne to their summit level north of Barrowford, with access thence over their line into Craven. (At this time the Leeds & Liverpool's route was run north-west of Colne, Nelson and Burnley by way of Padiham, and so north-east to Foulridge.) It would of course also have given them a route to Leeds alternative to that by the Calder & Hebble and Castleford—useful for bargaining—and given the Leeds & Liverpool a link with Manchester, considerably less roundabout, though hillier, than the one at Red Moss they were discussing with the Manchester, Bolton & Bury. The Leeds & Liverpool soon realized that the Rochdale had picked a useful line. They therefore worked to take over the potentially lucrative Worsthorne–Colne section, leaving the Rochdale the heavy lockage down to Todmorden, and proposed a compensation toll as well to discourage non-limestone traffic. By December they had agreed in principle to a junction near Colne, though now the Rochdale began to look, if only for propaganda purposes, for an independent source of limestone in the Lothersdale area north-east of Colne (see Chapter VII).

The promoters havered between a broad and a narrow canal. The Duke wanted it narrow, as he did the proposed Stockport and Lancaster canals, because then he would himself keep the carrying

17. The Pennine canals

trade to Liverpool. And, given that if it were to be broad, £350,000 would be needed, the promoters were also inclining that way. They decided to wait until after they had got their Act before choosing, but to seek subscriptions for the larger sum. William Crosley in a letter to Rennie commented sardonically that he expected 'very florid debates respecting wide and narrow canals'.[11]

In the welter of the various ambitions of the Bury & Sladen, Ashton (now looking for a junction with the Rochdale near Oldham), Leeds & Liverpool and the Duke, the promoters failed to realize that the opponents to be most feared were the millowners, deeply interested in avoiding any interference with their water power. They maintained that 59 mills, mostly industrial, would be affected, and foretold unemployment. The company retained Robert Mylne, great expert on water supply matters, to avoid his appearing against them, and Rennie proposed to supply the canal mainly by pumping, with three steam engines on the Yorkshire side and eight in Lancashire, at a cost of £3,301 p.a., and another at Burnley on the branch at £896 p.a. But the Bill was lost on an amendment moved in the millowners' interests. Rennie copied an epitaph into his notebook.

'On a fair and virtuous young maiden call'd the Rochdale Canal who was barbarously murder'd near Old Palace Yard, Westminster, on the night of 21 March 1792.

Here lie the remains of the Rochdale Canal
Put to death by the hands of a Cruel Cabal
Of the Buff and the Blue
And their desperate Crew
join'd by twenty or more
Who ne'er join'd them before.
Gentle reader, ah! pity the doleful condition
Of this maiden thus slain by this fierce coalition,
Yet rejoice, with her friends, in the pleasing reflection
That she died—full of hopes of a blest resurrection!'

The subscribers settled their accounts, but were determined to go on and get their canal, for as early as April John Royds was off to see the Duke. At a general meeting on 23 August, the room booked at an inn proved too small for all the former subscribers who came 'and the Meeting being intruded upon by persons who are not subscribers and consequently not within the Intention of the Meeting',[12] a move of subscribers only was made to a neighbouring schoolroom.

At this meeting the committee reported an approach from the

Bury & Sladen promoters, proposing that the Rochdale should abandon its own project and join with them. This idea was turned down, but the meeting agreed to go ahead with a Bill for a canal only from Chelburn beyond Littleborough past Rochdale and by Castleton to join the Bridgewater Canal near Cornbrook, the Bill to include the former branches and also reservoirs. A new committee was appointed, which included the energetic Dr Drake, vicar of Rochdale and John Gilbert, junior, a new subscription raised, no one being allowed to put up more than before, and a deputation despatched to meet the Duke and the millowners. A few days later, the committee realized that if they did not say they wanted to go on to Sowerby Bridge, the Bury & Sladen promoters would. They therefore hurriedly agreed to reinstate their old line onwards from Chelburn to frighten off the Bury & Sladen but to withdraw it at the last moment. The interview with the Duke went badly. He 'positively objected to a junction . . . at Corn-brook, or at any other place', but was willing to support a narrow canal from Manchester to Chelburn.

By September, Crosley, helped by John Longbotham, was surveying for reservoirs to catch rain and flood water without interfering with the Roch, Irk or Medlock. The Bury & Sladen then gave notice that they also proposed to extend to Sowerby Bridge, and the Rochdale supporters set themselves to seek accommodation with the Duke, the Calder & Hebble and the millowners in a somewhat ticklish situation.

Perhaps because of the Bury & Sladen's revival and so the potential threat to his own line, the Duke relented in October, to the extent of asking the enormous compensation toll of 3s 8d (18½p) a ton on all traffic except flagstones from the Rochdale, such 'Rate of Tonage . . . to be invariable for ever'.[13] Desperate to get his support against the Bury & Sladen, the Rochdale committee thought the terms 'reasonable and ought to be accepted', and Royds was asked to convey 'the grateful Thanks of this Committee to his Grace for his good Intentions'.[14] Mylne was again retained, and by promising to use gauges on their reservoirs and supply streams they hoped to reassure millowners that only surplus water would be taken. Fortunately, as the Bury & Sladen's prospects improved, so the Duke's terms came down to a compensation toll of 1s 2d (6p) a ton, said to be the amount already taken by him for warehousing and wharfage of goods awaiting transhipment to and from road, and which would be lost to him when the Rochdale was built.[15]

The new Bill was for the whole distance to Sowerby Bridge, with a 3,000 yd summit tunnel and eleven reservoirs. The Colne branch had been dropped, and no junction with the Ashton was provided for. Rennie's estimate was £291,929. He had decided in favour of narrow boats, and told the Parliamentary committee that '30 Boats of 25 tons each will be more than adequate to the Trade of the Country', which was rather to damn the proposal with faint praise. Strict water clauses were agreed to protect millowners on the Roch and Irk, while the choice of narrow boats was intended to reduce the canal's need for water. Prospects looked good to some, for in February shares were changing hands at 25 guineas,[16] but again the Bill was lost, this time by only one vote, in spite of the Duke's support. So, five days later, was that for the Bury & Sladen, by a much larger margin.

Once more the promoters began to plan. This time they worked even harder to reach an understanding with the millowners, and to convince them, by practical gauging of supplies in the Pennines, that the canal's reservoirs could be filled with surplus water without affecting their power. Their arguments were then sent to all members of Parliament. They also sought petitions in favour of the canal from much farther afield than before: from Hull, Bristol, Gloucester, Tewkesbury, Bewdley and the Potteries, among other places.

Every effort was indeed necessary, for not only was it likely that the Bury & Sladen Bill would be reintroduced, but there was now before the public a plan for another Pennine canal, the Huddersfield (see Chapter XII). This brought the Calder & Hebble in more enthusiastically on the Rochdale's side, and in October they agreed upon terms for a junction, whether the Rochdale were to be built broad or narrow. Finally, William Jessop was persuaded, though 'so much engaged',[17] to survey parts of the canal where the millowners were most concerned about water supplies, and to give evidence to the Parliamentary committee. His report was well publicized in the press.[18] Judging by his notebooks, Rennie seems to have had little to do with the third Bill, though the deposited plan was surveyed by William Crosley under his direction. The Bury & Sladen were also trying again. This time, on 4 April 1794, in a shower of supporting petitions the Rochdale got its Act.[19] So, on the same day, did the Huddersfield. The three applications had cost the shareholders over £12,000. The Bury & Sladen failed once more.

The Rochdale's line was to run from the Duke's Canal at Castlefield in Manchester via Failsworth, Littleborough, Hebden Bridge and Mytholmroyd to join the Calder & Hebble at Sowerby Bridge, with branches to Castleton (Rochdale) and Hollinwood (for Oldham). The Duke was authorized to receive a compensation toll of 1s 2d (6p) a ton on all goods except flagstones (2d) passing either way through the junction lock.* The Ashton were given powers to join the Rochdale without loss of water to the latter, and any waste water, either from the Ashton or the Rochdale, had to be passed to the Bridgewater Canal by the Duke's Bank Top tunnel (see p. 33). The 1794 clause was varied in the later 1800 Act to enable waste water off both canals to pass to the Bridgewater either by the Rochdale Canal or by the tunnel. At Sowerby Bridge the Calder & Hebble were to build warehouses and wharves for the Rochdale, for which the latter were to pay rent, the Rochdale company being prohibited from building either within 500 yd of Sowerby Bridge. Moderate tolls were enacted: 1½d a ton for coal, stone and minerals, 1d for lime, limestone, manure and sand (less if craft did not pass a lock).

The authorized capital was £291,900 in £100 shares, with power to raise £100,000 additionally. This was based on Rennie's estimate for the second Bill of 1793, when, as we have seen, he had a narrow canal in mind. But the Act now provided that the Rochdale's locks should be broad, the same size as those on the Bridgewater, so that craft could pass from the Mersey tideway on to the canal. The capital was therefore inadequate for the work that had to be done.

George Travis the archdeacon was given five shares bought with company funds in recognition of his efforts, was elected chairman, and asked to superintend the execution of the shareholders' plans. Contrary to a general impression, John Rennie had nothing at all to do either with the final layout of the canal or its construction: he does not appear again in the company's records after the Act had been passed, and indeed had played a small part in obtaining it compared with his work on the earlier Bills. It was William Jessop who, as soon as the Act had been passed, went over the proposed line that Crosley had staked out, and reported on 17 June 1794.[20] He proposed a canal 42 ft wide at surface (except in cuttings and on embankments) and 5 ft deep. Rennie's 3,000 yd summit tunnel† was eliminated, adding 14 locks but saving £20,000. He suggested the

* This was built and manned by the Duke, and remained Bridgewater property until 1887.

† Rennie had offered the alternative of 16 locks at the time of the 1793 Bill, but preferred the tunnel in order to ease his water supply problems.

position of each lock. Between Piccadilly, where the Ashton Canal would come in, and Castlefield, he foresaw a large trade, and planned locks of 7–8 ft fall. Elsewhere, except for one above Piccadilly, he hoped all could be given a uniform rise of 10 ft, to prevent waste of water and enable a spare gate to fit any lock. This ideal was indeed largely achieved in practice.

William Crosley and Henry Taylor were appointed resident engineers, though Taylor resigned almost at once, and Crosley carried on alone. Major questions were referred to Jessop, who paid the works a visit from time to time and wrote reports for the shareholders. Cutting began in 1794 with the most difficult sections, at the Sowerby Bridge end, on the summit, and between the Ashton Canal junction at Piccadilly and Castlefield, and was soon going on over most of the line east of Rochdale. Work then went steadily ahead, though in 1796 and later, in common with other companies, the Rochdale had to slow down construction to the speed at which it could get money in. One of the contractors was John Gilbert, son of that John Gilbert who had been the Duke's agent. In the spring of 1796 Jessop came to inspect what Crosley had so far done, and reported favourably: 'I must . . . say that, in my opinion, the works of masonry are in general sound and substantial, and that upon the whole there appears to have been good management in the conduct of the work, and economy in the expenditure.'[21] The proposed branch towards Oldham was not proceeded with, for the Ashton company had not only built theirs to Hollinwood, but had extended it nearer to Oldham by the private Werneth Coal company's branch 'under Colour of the Clause for taking Water into their Canal'.[22]

About the end of 1796, William Crosley senior died, to be succeeded by Thomas Bradley of Halifax and Thomas Townshend. A few weeks later, on 24 February 1797, Travis also died, and the company had lost two reliable and competent men of widely different abilities.

The canal was opened on 24 August 1798 to Todmorden, and on 21 December of the same year to Rochdale.

> 'The new canal from Sowerby-bridge to Rochdale was lately opened for business. The Travis yacht first crossed the head level, decorated with the Union flag, emblematical of the junction of the ports of Hull and Liverpool, with colours flying, music playing, attended by the Saville yacht, and thousands of spectators; a display of flags on the warehouses, and sounds of cannon, announced to the rejoicing neighbourhood the joyful

tidings, which in the evening were realized by the arrival of several vessels, laden with corn, timber, &c. &c. &c.'[23]

By 1798 also, reservoirs had been built at Hollingworth* (130 acres), Blackstone Edge (50 acres) and Chelburn (16 acres). Work on the Piccadilly–Castlefield section had, however, been delayed because of water difficulties, and this was probably not opened until late in 1799.

In early 1799 the committee started to encourage long-distance trade, advertising that they wanted terms from carriers for carrying between Hull and Rochdale by water, and by land between Rochdale and Manchester. They had one offer, from Edward Thompson, working with John Handley, both of Hull, to carry bale and manufactured goods as follows:

	Owner's risk			*Part carrier's risk*		
	£	s		£	s	
Manchester to Hull, per ton	2	5	(£2·25)	2	10	(£2·50)
Rochdale to Hull, do	1	15	(£1·75)	2	0	
Hull to Manchester, do	2	0		2	5	(£2·25)
Hull to Rochdale, do	1	10	(£1·50)	1	15	(£1·75)

Selby charges would be 2s 6d (12½p) a ton less. These were a good deal lower than rates advertised at the end of 1797 for land carriage between Manchester and Sowerby wharf, and water transport onwards: Manchester to Hull, £2 15s (£2·75); Hull to Manchester, £2 10s (£2·50).[24]

Thompson would start with two vessels, to provide a twice-weekly service in each direction, and put on more as needed. He would also arrange road transport between Rochdale and Manchester. The company agreed to send all goods by Thompson's boats that were not otherwise consigned. The Thompson service, taken over early in 1800 by Richard Milnes, does not seem to have worked, for in October the company were asking the Aire & Calder whether they could provide craft to work to and from Rochdale, leaving at regular intervals whether loaded or not. In December the Commercial Society of Manchester complained of 'negligence, irregularity and partiality in the conveyance of Goods',[25] and the Rev William Hassal, the company's agent, was sent to Hull to find out how a reliable service could be started. The answer was to do it themselves. Hassal therefore agreed with 'several owners of Sloops'[26] for the hire of their craft, and Thomp-

* It was not finished until 1800. A steam engine was installed to pump water up to a point whence it could run through a 4-mile feeder to the summit level.

son & Gilder of Hull were appointed agents for the service. The new arrangements were advertised in January 1801, at £2 a ton up and £2 5s (£2·25) down for goods in company's boats. Jointly with the Calder & Hebble, the company also tried to persuade goods then carried by land from Leeds to Manchester to transfer to water, and to hold off John Rooth, agent of the unfinished Huddersfield Canal, who was energetically seeking business. They made some toll reductions in aid of their policy. By February 1801, also, Aire & Calder boats were working over the canal, and in November 1802 that company were given a counting house at Rochdale. They seem to have withdrawn later, for in 1816 an effort was made to get them back. Hassal must have done good work as agent, for when he retired in 1811, he was given a present of £500.

Meanwhile the canal had to be completed. By mid-1799 it was clear that the authorized £291,900 would not be enough, and that more attractive securities than mortgages might have to be offered. The Duke was lobbied for support, and the Ashton given the hint that their junction would not be made until they had agreed not to oppose increased tolls. A new Act of 1800[27] granted powers to raise £100,000 by annuities or promissory notes, authorized deviations at Chadderton and Failsworth, increased maximum tolls on most commodities except coal, and allowed the ending of interest payments on calls, which had cost the company £27,000.

Meanwhile, relations with the Ashton company were awkward. There were arguments over the making and paving of the streets round the junction, which was to be done by both companies jointly, over a bridge the Rochdale thought the Ashton should build, and about tolls, but the Act defined what was to be done, and by April 1800 the Ashton committee had been told that they could make the junction and use the Rochdale line to Castlefield. In October the Manchester, Bolton & Bury company were suggesting a junction with the Rochdale above the third lock from Castlefield, partly to get access to the Bridgewater Canal. The Rochdale did not object in principle, but the plan went no further then.

Construction continued, and on 23 September the canal reached Hopwood, near Castleton, from Rochdale, and was opened by 'the Travis yacht attended by crowds of the neighbouring people'.[28] Late in 1802 Townshend left, and William Crosley junior was appointed to succeed him, still under Jessop's supervision.

By mid-1803 money was short. The committee sought another Act, and in the meantime had a drive on arrears, to keep work going and reduce their overdraft. The 1804 Act[29] authorized the raising of another £70,000. This was done by writing down the existing 2,861 shares, £120 called, to £60, and issuing a similar number of new £60 shares to existing shareholders at £20, and to new subscribers at prices from £50 to £64. The result was an ordinary capital of 5,722 £60 shares.

Trade was waiting. Charles McNiven of the Mersey & Irwell, writing from Manchester to Liverpool corporation on 9 September 1804, says:

'The Communication from Manchester to Hull through the Rochdale Canal will be opened this Winter . . . and I dare say you are aware of Timber, & other Goods having found their Way here from that Port during the Time it was only navigable to . . . Rochdale and under the disadvantage of twelve miles of Land Carriage.'[30]

On 21 December 1804 the canal was finished from Hopwood to Piccadilly, the first of the trans-Pennine canals to be completed. It was followed by the Huddersfield Canal in 1811, and the Leeds & Liverpool in 1816. The company had previously sent George Tindall in his boat *Mayflower* to pick up a cargo in London and come back in time for the opening ceremony, but sadly he was delayed by storms at sea on his outward journey, and got back late. Nevertheless:

> 'On Friday the Rochdale Canal . . . was opened in great style. The bells at Manchester commenced ringing at half past two, and the Company's passage boat and yacht, the Saville and Travis, were greeted from the banks for a great distance by an immense concourse of spectators, with many a vociferous cheer of grateful approbation. . . . The two vessels were filled with the gentlemen proprietors and their friends, attended by the band of the Fourth Class Volunteers, who continued to play many loyal and patriotic tunes. . . . Each gentleman, together with every servant and workman of the company, wore in his hat a blue ribbon, with the inscription of gold letters—"Success to the Rochdale Canal", and on the flag of the yacht was inscribed "Royal Rochdale Canal".'[31]

The company considered that they and their engineer had done a good job, and we may judge them right:

> 'Your committee flatter themselves . . . that, considering the magnitude of the works, and the difficulties of the undertaking, few canals have been prosecuted with more uninterrupted success. The masonry, and other parts of the work, have hitherto stood the test, with fewer misfortunes than are commonly ex-

perienced in such undertakings. And what adds to the *merit*, is, that few tracts of country have been so unfavourable as the greatest part of the line.'[32]

The Rochdale Canal stands with the Grand Junction Canal as among William Jessop's greatest works.

As built, the canal was 33 miles long, and rose by 56 locks (including the junction lock at Castlefield) to West Summit beyond Littleborough, then fell from Warland by 36 more to Sowerby Bridge. The ½-mile Rochdale (Castleton) branch was built, but not that to Hollinwood. There was a 40 yd tunnel at Sowerby Bridge,* and one of 336 yd at Deansgate.† A short cutting through hard rock near Sowerby was 50 ft deep at one point, and the summit cutting reached 38 ft. Through Manchester the canal had been side-walled for 4,560 yd. The Rochdale's locks, like those on the Bridgewater, took craft 74 ft × 14 ft 2 in. But those on the Calder & Hebble only took vessels up to 57 ft 6 in. long, though of the same width. Therefore cargoes carried in Bridgewater and Rochdale barges, or in full-length narrow boats, had to be transhipped at Sowerby Bridge. On the other hand, Calder & Hebble and Aire & Calder craft could work right through.

Years	*Tolls*	*Tonnages*	*Divs per £85 share*			
	£	*Tons*	*£*	*s*	*d*	
1806–08	13,187					
1809–11	17,666					
1812–14	19,352	223,032		13	4	(66½p)
1815–17	19,712	287,841	1	6	8	(£1·33½)
1818–20	22,658	322,848	1	13	4	(£1·66½)
1821–23	27,201	406,952	2	6	8	(£2·33½)
1824–26	37,593	487,819	4	0	0	
1827–29	36,794	498,402	4	0	0	

The company's finances were finally cleared up by a fourth Act[33] of 1806. Under this, calls of £25 were made on the shares to pay off debts and temporary loans of some £117,000. The final position was therefore 5,722 shares of £85 nominal, less 59 shares bought in by the company and cancelled, so that dividends were later paid on 5,663. The cost of the canal had probably been rather over £600,000. It was, however, an immediate success. Above are figures, averaged over three-year periods, until 1829. The opening

* Length as given in a canal report for 1804. It was presumably later lengthened to 43 yd, the length given in *Bradshaw*.

† Also called Knott Mill or Gaythorn. This tunnel was shortened and widened in stages to its present length of 78 yd.

To Bolton & Bury
MANCHESTER BOLTON & BURY CANAL
Salford
VICTORIA STATION
LNW
Surveyor of Salford's Wharf
Oldfield Road Coal Wharves
L&Y
Lock 6
Locks 4 & 5
New Yard Coal Wharf
Phoenix Wharf
Salford Tunnels
Old Botany Wharf
MANCHESTER
Stanley St. Coal Wharf
River Lock 1 & 2
Old Quay Yard
Dock Yard
Lock 1
Lock 2
Brunswick Wharf
Lock 3
LNW
Old Quay
Ordsall Lane Coal Wharf
MANCHESTER & SALFORD JUNCTION CANAL
Tunnel 499 yds.
LIVERPOOL RD. STATION
Lock 3
Lock 4
Reservoir
Pumping Engine
Stop Gate
Dickenson Stre
Arms
New Quay
Kenworthy's Wharves
MERSEY & IRWELL NAVIGATION
Merchant Co's Wharf
Junction Lock 92
Deansgate Tunnel
Lock 91
Stone Wharf
Lock 90
Lock 89
Victoria Quay
Hulme Lock
Hulme Junction
Slate Wharf
Grocers Wharf
Gathorne Bridge
Old Basin
New Basin
Castle Quay
Middle Basin
Knott Mill Packet Station
Hulme Hall Dry Dock
CASTLEFIELD WHARVES
To Warrington
Timber Yard
Dock
Commercial Mill Basin
Hulme
Cornbrook Bridge
General Wharf
BRIDGEWATER CANAL
Ashton's Bridge
MSJ&A
To Runcorn
Bridge
Lock
Double Lock (side by side)
Staircase Lock
General Wharf G
Coal Wharf C
Timber Wharf T
Stone Wharf S
Lime Wharf L
Huddersfield Canal Wharf HC
Peak Forest Canal Wharf PF
Anderton Co's Wharf A
NOTE -
RJD

18. Ro

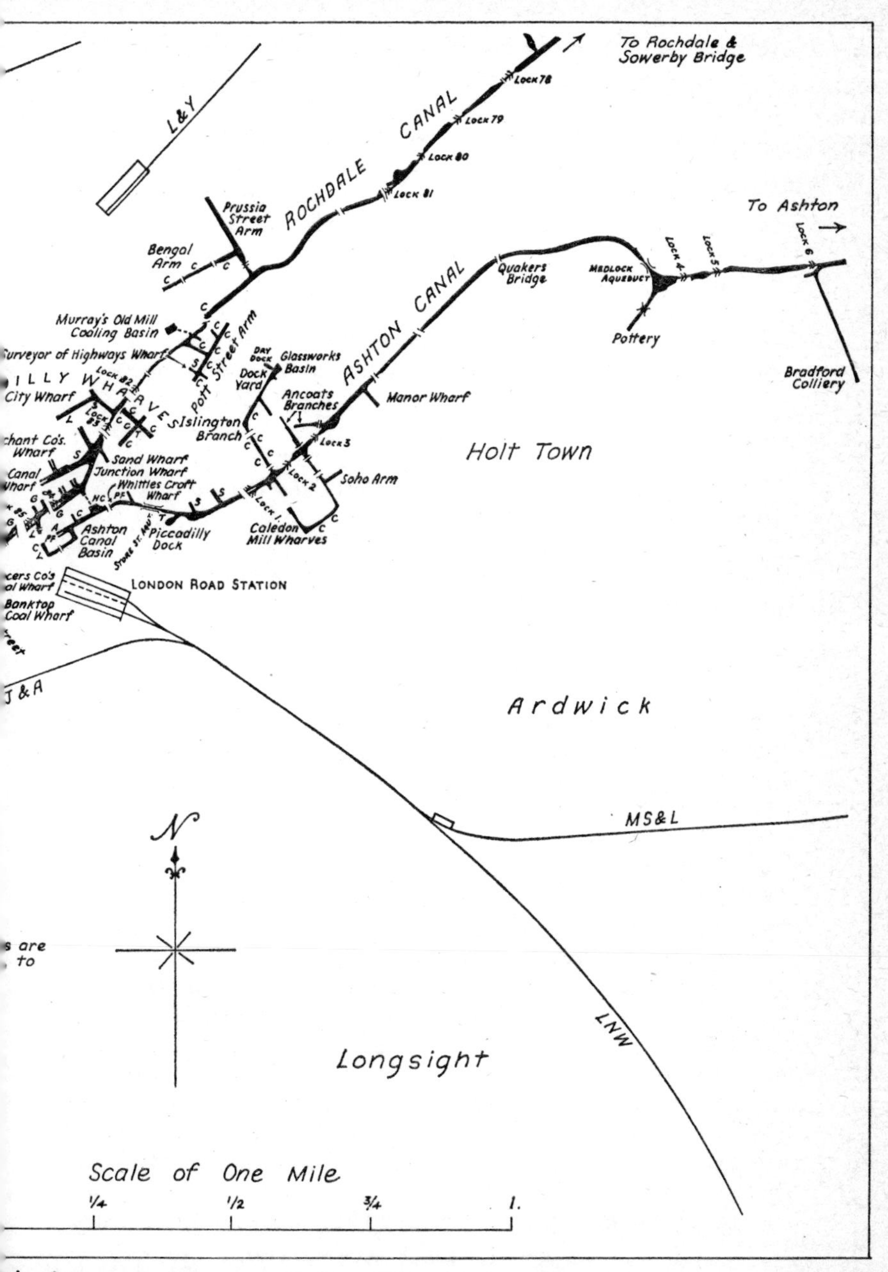
To Rochdale & Sowerby Bridge
L&Y
ROCHDALE CANAL
Lock 78
Lock 79
Lock 80
Lock 81
Prussia Street Arm
Bengal Arm
Murray's Old Mill Cooling Basin
Surveyor of Highways Wharf
Pott Street Arm
Dry Dock
Glassworks Basin
Dock Yard
Lock 82
City Wharf
Lock 83
Islington Branch
Sand Wharf
Junction Wharf
Whittles Croft Wharf
Ashton Canal Basin
Piccadilly Dock
Store St. Aqu.
Caledon Mill Wharves
Lock 1
Lock 2
Lock 3
Soho Arm
Ancoats Branches
Manor Wharf
ASHTON CANAL
Quakers Bridge
Medlock Aqueduct
Lock 4
Lock 5
Lock 6
Pottery
To Ashton
Bradford Colliery
Holt Town
LONDON ROAD STATION
Banktop Coal Wharf
J & A
Ardwick
MS&L
LNW
Longsight
N
Scale of One Mile
1/4
1/2
3/4
1.

in 1849

of the Huddersfield Canal in 1811 caused a brief fall in the takings, the result of toll-cutting, but they quickly recovered. That of the Leeds & Liverpool in 1816 had no perceptible effect.

The company's contract boats between Rochdale and Hull had been given up in February 1804, presumably because trade had begun to establish itself. They now set themselves steadily to attract traffic. A basin and wharves were to be made at Manchester, and a warehouse built. The wharfinger there was told that he

> 'must use his Exertions in laying himself out, and soliciting Goods, in Manchester; But that he is to attend to particular Consignments; and not use any partiality in sending off Goods'.[34]

Rivalry with the Huddersfield Canal led to some skulduggery as between agents and carriers, and early in 1806 the company minuted that

> 'all the agents who are detected in being employed, or receiving Emolument from any other Concern be discharged as soon as their accounts are made up'.[35]

Such rivalry led to another Act of 1807[36] which authorized the company to charge boats 1d (narrow boats in pairs ½d) per lock for what had earlier been minuted as

> 'paying extra persons to aid and assist in passing the Locks, in order to their greater safety and expedition in navigating, and for the preservation of the water'.[37]

In turn they bound themselves to pay 5s (25p), should a boat be delayed at a lock. A programme of building 30 lockhouses was then begun* and also of additional reservoirs to make sure that water supplies remained sufficient. These were Light Hazzles (30 acres) and Whiteholme (92 acres).

Having got trade started, the company agreed at the end of 1807 themselves to operate as carriers between Manchester and Rochdale with three specially built craft, and met the Aire & Calder to persuade them to work boats through to Manchester. By early 1809 company's boats were leaving Rochdale three times a week, and freight rates had been cut. In 1808 tolls were brought down. Agents and wharves were then chosen at Liverpool and Gainsborough on the Trent to receive and ship goods for the Rochdale Canal. A useful way of encouraging trade was to offer free warehousing or wharfage. In 1809 the company made no charge for 14 days on all goods except timber, and for two months on stone, flagstones and slate.

In 1810 the company appointed a superintendent, John Crossley,

* Twenty-six had been built by May 1809.

already a member of the managing committee, at £300* per annum, mainly to encourage trade, keep an eye on agents, and take money. They found it wise in promoting business to keep out of the way of the Bridgewater trustees, and minuted in 1810 that

> 'the Canal Company should not interfere with Mr. Bradshaw's carrying concern, by shipping Goods from the Company's Wharf at Liverpool for Manchester'.[38]

Nevertheless, a directory of 1815 said:

> 'The Rochdale Canal Company have provided a basin and extension wharf, with suitable accommodation, near the King's Dock, Liverpool, for such vessels as trade thither by way of the Rochdale Canal.'[39]

In 1811 they retired from carrying, selling their craft to a carrier, Job Cogswell, who undertook to continue the business.

A sign of competition with the newly opened Huddersfield Canal were the terms of a warehouse lease at Manchester offered to the Grocers' Company: that they should not carry goods on the Huddersfield Canal going to, or coming from, points beyond Huddersfield. Another was the drawback of 2s (10p) a ton on the more expensive goods carried to and from Wakefield, Leeds or beyond, and later extended to places on the Calder & Hebble and all points on the Leeds & Liverpool, Barnsley and Dearne & Dove canals.

We get a picture of which the canal was carrying from the tonnage figures for 1812 and 1819:

	1812 *tons*	*1819* *tons*
Corn	20,375	40,553
Coal	42,509	95,470
Stone	26,033	45,255
Lime	11,735	13,458
Wool	3,070	4,452
Timber	3,186	6,270
Salt	4,127	2,820
Merchandise	5,793	21,122
Sundries	82,795	87,650
	199,623	317,050

These figures can be compared with the 38,899 tons carried on the narrow-boat Huddersfield Canal in 1817:

* Raised to £500 per annum in 1817.

	tons
Corn	1,562
Coal	10,823
Lime and limestone	2,027
Stone and bricks	9,254
Pig iron	525
Merchandise	14,708
	38,899

The growth in the corn trade is notable. The opening of the Rochdale Canal encouraged corn from the Gainsborough and Lincoln areas to cross the Pennines to feed Lancashire workers. Previously, much had come from western ports to Liverpool, and in February 1807 the Mersey & Irwell company noted 'a considerable diminution of corn carrying on their navigation because a large part of the supply of the neighbourhood was now brought by the Rochdale Canal'.[40] Later, the westwards trade in flour, malt and bran was encouraged by drawbacks, but one raises an eyebrow at the specially cheap toll of ½d a ton for stable dung, street sweepings and nightsoil from Manchester in the same returning boats.

Timber came in at Hull for a long haul to the west. A good deal of the coal trade, that on the Lancashire side, was local: on the other, Yorkshire coal came up naturally as far as Littleborough, but had to be given a drawback to penetrate farther. A notable item, both on the Rochdale and the Huddersfield, is the merchandise figure. This, probably largely finished goods, must have been Lancashire exports going to Yorkshire and for shipment from Hull coastwise to London and elsewhere, and also abroad, and also groceries etc coming in, though much of the latter was probably classified under sundries. Finally, as can be seen, the Rochdale's effort to get the salt traffic to Hull away from the Trent route had not been successful. Corn and merchandise were the lucrative traffics: in 1818, out of a toll revenue of £23,898, £7,555 came from corn and £3,042 from merchandise.

In 1811 a first dividend of £1 per £85 share was paid. Rate cutting and capital expenditure on reservoirs, warehouses and other facilities did not permit a higher rate until 1821, when £4 was declared and maintained unbroken to 1835. It paid to have warehouses. In 1813 room was lacking, and some vessels had to go to the Bridgewater Canal to unload. Notice was therefore given to the Grocers' Company, who did not use the Rochdale Canal (they traded mainly on the Bridgewater), a new grain warehouse was ordered for Rochdale, and negotiations opened with the Calder & Hebble for spare accommodation. The period of free wharfage and warehousing was extended in 1815 to 21 days for grain and merchandise, 42 days for stone and slate.

Post-war depression caused one carrier to fail owing £525, and there were difficulties in providing a through carrying service. In May 1816 the committee were recommended by the shareholders to try to get a certain and regular conveyance of goods 'by an Establishment of Vessels from port to port, between Liverpool and Hull'.[41] Later in the year they proposed to ask the Aire & Calder 'how far it is probable their vessels might be induced to come to Rochdale, or Sowerby, and to Sail at stated periods, by offering them accommodation there'.[42] Then, in the spring of 1817, the company decided itself to return to the carrying business, by contracting for craft for charter up to seven years, to work between Liverpool and Wakefield. This plan seems then to have been dropped, perhaps because independent carrying improved, but they got an offer from an independent group headed by Isaac Leach if they would increase the long-distance drawback from 2s (10p) to 3s (15p), and give free warehousing, to run a regular service between Lincoln and the Trent & Mersey Canal via the Rochdale. They accepted the offer.

At this time the three main wharf and warehouse centres were at Piccadilly, Manchester, where the wharfinger got £340 per annum and a house, Rochdale (£220 and a house), and Sowerby Bridge (£150 and a house); wharfingers had to provide their own clerks out of these sums. There were warehouses also at Gauxholme and Todmorden. The clerk at the tunnel lock where the Rochdale joined Castlefield basin got £110 per annum and a house.

By 1820 Job Cogswell, the carrier who had earlier bought the company's boats working between Rochdale and Manchester, was willing to work two boats a week regularly to Wakefield, if the company would allow him £100 to pay for an agent there. They agreed.* In 1822 he put on a service of three vessels carrying bale goods between Halifax and Liverpool, but working from the company's basin at Sowerby Bridge. But he died in 1824, though his firm carried on.

Business was good, and in November 1822 the committee

* The company also helped to pay the agent at Wakefield of the Rochdale & Halifax Merchants Co in 1826. Carriers were also helped by lending them money to start in business, or to build boats, and by providing stables for their horses.

ordered: 'That a Gateway built of Stone rusticated,* be erected at the Entrance into the Manchester Wharf, between the office and the new Warehouse there.'[43] This sign of confidence was almost immediately followed by letters sent to the treasurer about the proposed Liverpool & Manchester Railway. He was told to reply by saying 'that the Committee do not view the Measure with much alarm'.[44]

All the same, the company took precautions against the spread of competition, whether from railways, from a Stockport branch of the Bridgewater that would have by-passed their canal between Castlefield and the Ashton junction, or from the Leeds & Liverpool now that the Leigh branch was open. On the one hand, they closed down on information upon how the company was doing: 'That no report be made by the Committee to the proprietors as has heretofore been usually done, but that the Treasurer remit the dividends without any explanation, remark or comment, regarding the present state of the Canal or its pecuniary concerns.'[45] On the other, they worked to improve facilities: a new warehouse at Rochdale; new wharves and cranes where they were wanted by those who found public wharves inconvenient; Sunday opening of the tunnel lock at Castlefield and, unfortunately for the staff, 'in case of need, the Wharfingers shall stop up at Nights to render every facility for the despatch of all Vessels which are under the necessity of sailing during the night'.[46]

Carriers were at this time allotted space in the company's warehouses proportionate to the business they did on the canal. There were six accommodated at Manchester in 1825:

	rent £
Rochdale & Halifax Merchants Co	200
Manchester & Liverpool Union Co	140
John Thompson & Co	100
Barnaby Faulkner, Reeder & Co	80
James Veevers	30
George Thornton	25

In 1823 the committee began to take an interest in a possible trade with Sheffield via the Barnsley Canal, influenced possibly by contemporary schemes for linking Sheffield with the Peak Forest Canal by a waterway across the Peak District (see Chapter XI), and early in 1824 offered 1s (5p) drawback on all Sheffield and Rotherham goods to and from Manchester which then paid a 3d per mile

* It can still be seen.

toll, if the Calder & Hebble, Barnsley, Dearne & Dove and Sheffield companies would also make reductions.

Business steadily increased. Small toll reductions, or grants of drawbacks, were being made all the time to encourage special traffics, many in collaboration with the Calder & Hebble, with whom the Rochdale were working more and more closely: in June 1824, the Calder & Hebble directors were asked to dinner, and to see the new Rochdale warehouse.

Fly-boats (now first mentioned), keels and bale vessels running to timetable (whether laden or not) were in 1825 given day and night priority if advance notice were given. Gas lighting was now put into the Manchester wharves and offices, and facilities provided there for gauging boats.

In early 1827 they saw a plan of a projected independent canal from Wakefield to Ferrybridge, which they decided strongly to support as an improved line to Hull. The Aire & Calder reacted with their own alternative, but proposing to charge their old distance tolls on the shorter new line. The Rochdale decided to argue with them over this. Meanwhile, they conceded a toll reduction on the important malt traffic from Goole and Selby.

The company had been accustomed to borrow the Calder & Hebble's yacht *Savile* for their periodical committee inspections of the canal. In 1832 they decided to have a 'sufficient yacht'[47] of their own. It was built of wood by Smith & Marrow of Liverpool for £320, and the committee meeting of 9 July 1833 was proudly datelined from *The Rochdale*.

In 1829 one of the company's Manchester warehouses was burnt down, and a member of their staff lost his life. This woke them up to the need for insurance on all their properties, and caused a spate of bye-laws: no one to use unguarded candles or lamps in a warehouse; no boat with an unguarded light or fire, or with anyone smoking, to come nearer to the warehouse than a boat's length; no smoking in warehouses. Soon afterwards they ordered gas to be installed in the two Rochdale warehouses, and iron fire doors on each storey of some Manchester ones.

In October 1830 a railway was proposed from Manchester to Sowerby Bridge. This was surveyed by George Stephenson, who was appointed engineer with James Walker, a line by Littleborough and Todmorden was chosen, paralleling the canal, and a Bill prepared. The canal company started by writing to other canal concerns to ask whether they would be willing to unite in opposing 'all or any of the projected Schemes for making Railways', and suggest a meeting.[48] This was indeed held in London, but seems to have been inconclusive. When the committee saw the Bill, they decided that it should 'be opposed in every possible way'.[49] A deputation of committeemen and the engineer was to go to London to do so. The latter was also ordered to re-survey the railway's proposed line and estimate its cost of construction and maintenance, while William Crosley was asked to check on the actual traffic on the turnpike road between Manchester and Sowerby Bridge. The Bill was defeated. In 1834, still free from competition, they gave R. Cort £20 for 'Services rendered to the Interests of Canals' by his pamphlet, *Railway Impositions Detected.* [50]

In August 1831 all drawbacks were cancelled, and a new set prepared. These give an interesting list of the traffics the canal company were anxious to encourage:

Hull to Manchester—deals, timber;

Sowerby Bridge to Manchester—coal, corn for manufacturing, merchandise;

Luddendenfoot to Bridgewater Canal—stone and flagstones;

Burnley (road transport to Todmorden) to Manchester—cargoes paying 3d a mile (e.g. bale goods);

Rochdale to Manchester—manufactured cotton and woollen goods, flour, meat and bran;

Manchester to the Aire & Calder—salt;

Manchester to Sowerby Bridge—blue slate;

Manchester to Rochdale—cotton and wool.

In the same month the Peak Forest company gave a drawback on lime traffic carried to the Rochdale Canal.

In August 1832 the committee were offering for land to build a branch canal to Heywood, and then asked for estimates of cost and traffic. A special shareholders' meeting of 3 August 1833 authorized it, to be built without an Act. Work started, a warehouse was put in hand at Heywood, and on 10 April 1834 the 1½ mile level branch was opened by the committee in *The Rochdale*.

The company's prosperity was growing quickly. The Liverpool & Manchester Railway, and improvements on the water lines, brought more traffic from Liverpool on to the canal, while that from east of the Pennines increased correspondingly. Tonnage, 500,559 tons in 1829, was 875,436 in 1839; toll receipts were up from £38,147 to £62,712, and the dividend per £85 share from £4 to £6. Improvements were made: reservoir extensions; a new warehouse at Manchester; walling and widening of canal sections where craft could not pass; and in 1836 their principal agent was

Page 285 Manchester, Bolton & Bury Canal: (*above*) the Clifton aqueduct, on the left, carries the canal over the Irwell. On the right is Fletcher's Canal, and in the background the Clifton railway viaduct; (*below*) Prestolee aqueduct over the River Irwell. The line to Manchester is to the left

Page 286 Rochdale Canal: (*above*) *Shamrock* and *Peter* frozen in at Walk Mill, Chadderton, in February 1905; (*below*) *Mabel* passing *Ellen* at Todmorden, *c.* 1890

sent, with the company's chairman and three committeemen, to see the Birmingham Canal Navigations 'with the View of gaining Information as to the mode of improving Canals'. On the other hand, the proposed Manchester & Salford Junction Canal (see p. 126), to connect the Rochdale Canal to the Mersey & Irwell, met with little enthusiasm. Satisfied that they would not lose water, they let it go ahead, but did nothing to help it forward.

They were indeed preoccupied with the revival in 1835 of the Manchester–Sowerby Bridge railway project, now as the Manchester & Leeds. At first they opposed; then the railway projectors asked the canal company to receive a deputation, and agreement was reached, as also with the Calder & Hebble. The Bill went through, but the Rochdale obtained protection on railway crossings of the canal, and were sufficiently satisfied to congratulate the deputation they had sent to London. In 1837 they decided to oppose 'the connexion Railroad* in Manchester by all means in their power'[51] in the interests of their through trade with Liverpool. In that year also the Manchester & Leeds Railway offered to buy the Heywood branch, since they proposed to build one themselves, but were refused.

The main-line railway opened from Manchester to Littleborough in July 1839, after which a passenger packet boat ran for a time between the station at Bluepits (Castleton), 1½ miles away, and Heywood canal basin. It seems to have been taken off at the end of 1840, before the branch route was opened in April 1841.

In September 1840 the Manchester & Leeds met the principal water carriers on the Manchester and Hull line, and told them they had

> 'no wish to become carriers themselves provided the Public Carriers would take up the business with proper spirit. The price named was 4d per Ton per mile†—the Carriers to state what discount they would require on certain articles on the carriage of which the competition with the Canals would be the greatest—the carriers expressed their satisfaction with the Company's terms and general arrangements and expressed a desire to remove their business to the Railway as soon as it was opened throughout'.[52]

The railway worked hard on the carriers. In September 1840 the

* The link between the Liverpool & Manchester and the Manchester & Leeds, made in 1844 when they met at Victoria station.

† This sum included locomotive power, waggons and tolls. Collecting, loading, sheeting, unloading and delivery to be paid for by the carriers.

JOHN BIBBINGTON,

LIME AND LIME-STONE MERCHANT,

CANAL WHARF, ROCHDALE,

AND AT THE

CANAL WHARF, HEYWOOD;

CANAL WHARF, BURY,

AND

MATHER STREET, MANCHESTER.

DEALER IN

IRONFOUNDERS' ROAD SAND,

FOR MOULDING AND FURNACES,

AND

LIME BURNER

AT DOVE HOLES,

CHAPEL-EN-LE-FRITH,

NEAR BUXTON,

DERBYSHIRE.

CARRIER BY WATER

FROM

ROCHDALE AND HEYWOOD

To MANCHESTER, STOCKPORT, ASHTON-UNDER-LYNE
STALYBRIDGE, HYDE, MARPLE, BUGSWORTH, and
WHAL[illegible]Y.

HEAD OFFICE, ROCHDALE.

19. A lime and limestone merchant's directory advertisement of 1851

managing director reported an interview with one of them, Barnaby Faulkner, who 'stated that he had made up his mind to come on the Railway and if one Carrier came they all would'. The former suggested to the board that carriers should only get a discount off the 4d if they brought a certain amount of their business with them.[53] In October it was Pickford's, 'who expressed their determination to send their London Traffic by an easterly route, and hoped the Company would shortly be prepared to receive their business on the Railway';[54] a draft agreement was agreed by the board in December, when letters were read 'from various Carriers expressing a wish to make use of the Railway as soon as it was ready for them'.[55]

In March 1841 canal competition was showing results, for Capt J. M. Laws, the general manager, reported to the railway board a falling off in carriers' traffic to Wakefield and Hull, and recommended cuts in cotton, grain and flour. At that date the railway company itself became carriers of cotton twist. In May they had talks with the carriers on sharing traffic, and agreement was reached on the Hull trade with some of them—Thompson, McKay, Faulkner, Buckley & Kershaw, and F. & W. Marsden.[56]

The railway had been opened between Normanton (for Leeds) and Hebden Bridge on 5 October 1840, and wholly on 1 March 1841; there was now rail competition with water carriage through to Hull. From the beginning of 1841 the canal company reduced its maximum toll to $1\frac{7}{8}$d per mile. Before that, they had tried to persuade the Calder & Hebble and Aire & Calder to join them in a concerted reduction, but without success. Now the Calder & Hebble did reduce $\frac{1}{2}$d a ton, and the Rochdale another $\frac{3}{8}$d. In January 1841 a deputation of Hull carriers asked for a further reduction to enable them to compete with the railway carriers, and the company agreed to press the Calder & Hebble, who agreed to reduce the rate for manufactured goods from Manchester to Hull. Joseph Priestley of the Aire & Calder explained that they had cut their highest rate from Goole from 8s $8\frac{3}{4}$d (44p) some years before, to 3s 1d ($15\frac{1}{2}$p), and from Selby from 7s (35p) to 2s $7\frac{3}{4}$d ($13\frac{1}{2}$p). He added: 'It is to be hoped that a large reduction by Water made simultaneously will frustrate the designs of the Railway Co.'[57]

The carriers now took a hand in cutting freights:

> 'You are no doubt aware of what has been going on with the Railway people and the Carriers within the last few days—They have followed us to 14s a Ton and we have now resolved to put it down to 10s and with the assistance of the Canals

continue the contest until they are brought to their senses. . . . We have only to add that we hope we shall not be left to fight the Battle as in our feeble hands it must end in disgrace and ruin. It must become a Canal question.'[58]

The canal companies then reduced again, but not enough for the Rochdale. By September maximum tolls on both the Rochdale and the Calder & Hebble were 1d per ton per mile. The pace was too hot for a newly built railway, and that same month Capt Laws of the Manchester & Leeds sounded the Rochdale on the likelihood of a rates agreement with the water companies. The Rochdale, whose share price was running at £40 against £150 two years before,[59] were agreeable, and in January 1843 the Aire & Calder and the Calder & Hebble also consented to meet the railway, though they wanted proposals from it first. They must have proved sticky over this point, for in February the Rochdale told the Calder & Hebble

'that this Committee are of Opinion that the period has arrived when arrangements should be made by the Railway and Water Companies to terminate the present injurious contest, that this Company is desirous to continue to act in cooperation with the other Water Companies, but they cannot recognize the propriety of permitting a private difference between the Calder & Hebble and the Railway Companies to obstruct arrangements required by the general interest. That both the Calder & Hebble and the Aire & Calder Companies be informed that if they are not prepared to enter into immediate arrangements with the Railway Companies, this Committee will feel at perfect liberty separately to negotiate for the individual interests of the Rochdale Canal Company'.[60]

This brought common action, and the water companies (except the Aire & Calder) met the Manchester & Leeds during February, and were told that the latter would follow them in an increase or a reduction. The Rochdale then moved its tolls upwards, and received a letter from the railway on 4 March saying: 'On the faith of your letter we shall raise our rates to the extent of ½d per ton per mile between Manchester and Sowerby Bridge on such goods as are included in your advance.'[61] Pressure from railway competition had been worst on coal and corn carrying. In the case of the latter, indeed, the Manchester & Leeds and Leeds & Selby companies had provided boats and tugs to carry Lincolnshire corn to their railways for Manchester.

Friendly relations having been established, the canal committee in March 1843 decided to offer the railway a 14 year lease at £40,000 p.a., themselves still to manage the canal. This idea was taken no further at that time, but agreement was on, even to the extent of the two concerns agreeing to act jointly to meet the competition of the Huddersfield Canal. The railway and canal carriers on the Hull route also reached agreement in May 1843, but had some difficulty in keeping it. In July, for instance, Pickford's on the railway were accused of undercutting the agreed rate on bales to Hull. By early in 1844, however, the railway company had moved towards buying the Huddersfield Canal and had leased the Calder & Hebble, though that concern eventually threw in its lot with the London & North Western (see p. 335). Rather remarkably, this agreement was never presented to a Calder & Hebble shareholders' meeting. The Rochdale was boxed in.

The Aire & Calder strongly disapproved of the take-over of the Calder & Hebble, complaining of a

> 'monopoly created by that Company* on goods carried by Railway and by water and the various acts of annoyance and oppression by which they seek to drive the traffic from the water to the Railway'.[62]

In October 1844 they complained to the Board of Trade.

In the meantime the Rochdale and the Aire & Calder jointly arranged heavy cuts in flour and timber tolls between Hull and Manchester—in the case of the Rochdale, a toll cut from 5s (25p) to 1s (5p). The Rochdale agreed that their manager should

> 'tell the Water Carriers that the Rochdale Canal Company are determined to have a fair share of business and are prepared to make any necessary sacrifice to retain it'.[63]

The result, perhaps helped by the Aire & Calder's threatened action over the Calder & Hebble, was another approach from the railway for a meeting 'to put an end to the opposition which is so prejudicial to the interests of both parties',[64] to which the Rochdale agreed.

The railway cautiously suggested a pooling agreement covering traffic that could be carried by either means, similar to that in operation between the Liverpool & Manchester Railway and the Bolton line. Because of a possible move by the Aire & Calder, they expressed themselves very carefully:

> 'The Directors declared that they wished at this peculiar time (not knowing what matter might be raked up against them in proceeding elsewhere) to speak with reserve on anything like a

* The Manchester & Leeds Railway.

> raising of dues to the public. The suggestion should be deemed a confidential communication . . . rather as a floating notion . . . than as a proposition.'[65]

Soon afterwards the Aire & Calder and Ashton companies agreed to the railway Bill for taking over the Huddersfield Canal, on condition of a clause giving the Board of Trade the right to intervene if prejudicial tolls were charged.

It was the year of the railway mania, and in August, like many of the canal companies, the Rochdale minuted that an engineer of eminence should be employed to 'survey the line of Canal and ascertain the practicability of converting the whole or any part of it into a Railway'[66] probably with the idea of expediting railway agreement.

The survey was done by James Thomson, who seems to have recommended against conversion, but suggested instead a new canal to replace the Calder & Hebble: the Rochdale therefore asked the Aire & Calder whether they would be willing to have a detailed survey done at joint expense. The Aire & Calder thought the plan immature, and would not contribute, though they

> 'expressed themselves desirous of adopting some measures to break or defeat the combination of the Calder & Hebble Navigation and the Manchester & Leeds Railway Companies'.[67]

So the Rochdale got their own engineer to do it. He reported that he

> 'had found the Country intricately involved in roads, rivers, canals and railways—that to form a new line of canal there would be at any time a task of much difficulty',

and that it could not be done in time for the new session.

Meanwhile talks had been going on between the Rochdale and the Manchester & Leeds Company, who suggested either amalgamation, or purchase, or a guarantee in perpetuity of canal dividends. In January 1846 the railway offered a dividend of £4 per £85 share in perpetuity, subject to an Act. By March, both sets of shareholders had agreed. The Bill was introduced in 1847, but was defeated by the Aire & Calder's opposition. (*To continue the history of the Rochdale Canal, turn to p.* 430.)

CHAPTER XI

To Ashton and the Peak

THE waterways of the north, in Lancashire and Yorkshire alike, were built to take barges or sailing flats. The narrow-boat canal, developed by James Brindley from the Worsley levels and mine craft, spread throughout the Midlands and into the south of England. But in the north it was represented only by one small system built by three closely associated companies, the Ashton-under-Lyne, Peak Forest and Huddersfield. It began in 1791, and by 1811 extended from the Rochdale Canal in Manchester to Ashton-under-Lyne and over the Pennines to Huddersfield, where it joined Sir John Ramsden's broad canal that led to the Calder & Hebble.[1] From Dukinfield near Ashton a branch, the Peak Forest, ran past Hyde and Marple to Bugsworth and Whaley Bridge, to tap the limestone of the Peak. In 1831 this group acquired new links to narrow canals: at Marple the Peak Forest was joined by the Macclesfield Canal[2] to the narrow-boat section of the Trent & Mersey at Kidsgrove near Stoke-on-Trent, and at Whaley Bridge by the Cromford & High Peak Railway to the Cromford Canal.[3] As a Pennine transport route the Huddersfield Canal was never of real importance; but the Peak Forest and the Ashton were in their time major carriers of limestone, lime and coal. By 1846, however, all three were railway-controlled.

Ashton-under-Lyne Canal

On 30 August 1791 notice of a Bill for the proposed Rochdale Canal appeared, which included a branch to or near Oldham from Manchester via Newton, Moston, Failsworth and Chadderton.[4] A fortnight later, and presumably as a consequence, another notice was published, announcing a parliamentary petition for an independent Manchester to Ashton-under-Lyne and Oldham Canal,[5] and a meeting as soon as particulars were ready.[6] This was held in

October, and the idea was well enough received for £21,900 to be subscribed on the spot.[7] By November the amount needed had been raised, and plans and estimates were being prepared.[8] These seem to have been done by Thomas Brown, an assistant of Benjamin Outram.

The Company of Proprietors of the Canal Navigation from Manchester to or near Ashton-under-Lyne and Oldham were authorized in June 1792[9] to build what was then an isolated line* of narrow canal from Piccadilly, Manchester, to near Dukinfield bridge, Ashton, with a branch to New Mill near Oldham. This branch was intended to run along the south bank of the River Medlock (with a short branch crossing the river there at Waterhouses) to Park Bridge, south-east of Oldham. Later the Fairbottom branch was built on the north side. The Act also provided for a short spur about ½ mile short of Ashton to run by the Dukinfield aqueduct to the far side of the Tame, and a steam engine to pump water. This spur was later to be used for a connection to the Peak Forest Canal. Waste water was to be conveyed at the Ashton's expense into the Duke of Bridgewater's Bank Top tunnel. The company could raise £60,000, and £30,000 more if necessary.

At that time the Ashton–Oldham area was an important producer of coal. Aikin says:

> 'The supply of coals of Manchester is chiefly derived from the pits about Oldham, Ashton, Dukinfield, Hyde, Newton, Denton etc. . . . The supply from the Duke of Bridgewater's pits at Worsley is less considerable although a very useful addition for the poor.'[10]

The Act therefore gave mineowners powers to build branch canals up to 4 miles long.

In March 1793 another Act[11] authorized a branch from Clayton junction to Heaton Norris, Stockport, through Openshaw, Gorton and Reddish, with a branch from it at Reddish along the edge of the Tame valley to Beat Bank, Denton, on the Stockport–Ashton road, for access to collieries in Haughton Green, and another branch from the Waterhouses aqueduct to Hollinwood. Another £30,000 could now be raised. So Stockport was at last to be given canal transport, after both the Duke's branch canal of 1766 and the project of 1790 on a similar line had failed. [12]

In September 1793 the company were advertising for an engineer,[13] but do not seem to have found one, for in August 1798 they

* The Rochdale Canal, with which it later connected at Piccadilly, was not authorized until 1794, though its first Bill was introduced in 1792.

minuted that the 'Works of the . . . Canal have been in many instances improperly managed for want of the assistance of a proper Engineer'.[14] James Meadows their agent had probably taken charge of construction. The line was completed between Ancoats, Manchester, and Ashton and along the Hollinwood branch, about the end of 1796, and opened to Stockport in January 1797.[15] The Fairbottom branch, which had not been separately authorized, was also opened in 1797.

The 1793 Act authorized a water supply feeder from the end of the Hollinwood branch. This was incorporated in a private canal about a mile long built by the Werneth Colliery company, which extended the Hollinwood branch to Old Lane colliery about a mile from Oldham. It was built under an agreement of 30 July 1795, which provided that the Werneth company should transport all their coal by canal, and make their surplus water available. In return, the Ashton company would allow them a drawback on tolls of 6d a ton in consideration of the costs of supplying the water. The Werneth Canal must have been opened soon afterwards. The Rochdale Canal company, whose 1794 Act included a branch to Hollinwood to tap Oldham's trade, considered that it had been tricked by this way of extending the Ashton beyond its authorized limit, and sought counsel's opinion upon 'whether the Ashton Canal Company, under Colour of the Clause for taking Water into their Canal, can extend the Navigation towards Oldham as they are now doing'.[16] Presumably they were told that the extension was legal under the clause allowing mineowners to build branches. The Werneth group of collieries had been built up by John Evans, William Jones and John and Joseph Lees, John Lees and William Jones being Ashton Canal shareholders. One may think, therefore, that the Rochdale company had reason for their complaint. Perhaps intending to regularize the position, the Ashton company gave notice in September 1797 that they would seek powers to take over the Werneth Canal,[17] but seem not to have taken the proposal further.

Having made sure of good supplies of coal, the canal company decided to stop work on the 3 mile long Beat Bank branch with its difficult clay slope, representing to William Hulton, who owned collieries at Denton that it was to have served, that they could not afford to finish it. In 1798 they introduced a Bill to allow them to abandon the branch after compensation to landowners for damage done, and raise another £30,000 (half of which would go to pay off existing debts) by mortgage, short-term or convertible notes, even though not all the £120,000 already authorized had been subscribed. When Hulton opposed the Bill, they offered to give him the unfinished branch. He refused, hoping to get the abandonment clause defeated.[18] But he failed, and the Act[19] passed. The Ashton company then told Meadows to help the Werneth Colliery company in selling the coal that was carried by canal. Soon afterwards they were allotted wharves at Stockport, Ancoats and the intended junction at Piccadilly, and in 1802 were building a warehouse at Hollinwood.

In July 1798 the Ashton company ordered warehouses to be built at Piccadilly on Lord Ducie's land, and at the end of the Stockport branch. They were desperately short of money, having by February 1797 called £135 17s (£135·85) per share;* and were then trying to borrow £20,000 from shareholders and on promissory notes. Their long-term prospects had, however, improved by the passing of Acts in 1794 for the Peak Forest Canal and the Huddersfield, both to join their own line, the former at the far side of their Dukinfield aqueduct, the latter end-on at Ashton. All three companies had many shareholders and committeemen in common, and shared the same Altrincham firm of solicitors, the Worthingtons.

The Peak Forest's line was opened on 1 May 1800 (see p. 309). By February 1798 the Huddersfield was almost finished from its junction with the Ashton to Woolroad (Saddleworth) and from Marsden to Huddersfield, though a great deal still remained to be done on Standedge tunnel. But trade could start, locally and by land carriage over Standedge, and in the same month the Huddersfield committee sent two delegates to the Ashton's general meeting to press them to complete their canal from Ancoats to its junction with the Rochdale, adding that the Huddersfield was set on foot in expectation of its being part of a through route between Liverpool and Hull.[20]

The Ashton company replied that they would seek powers to raise additional money to complete their line and 'have also determined that the said Canal Navigation shall be forthwith completed to the intended Bason at Piccadilly'.[21] By July they were building it to Piccadilly, where they and the Rochdale company had bought a large area of land from Lord Ducie, with Benjamin Outram as their contractor. Public wharves were to be laid out there, a warehouse built, and cranes provided. In August, however, they were

* In September they decided to square shares up to £100, by asking shareholders to even their holdings.

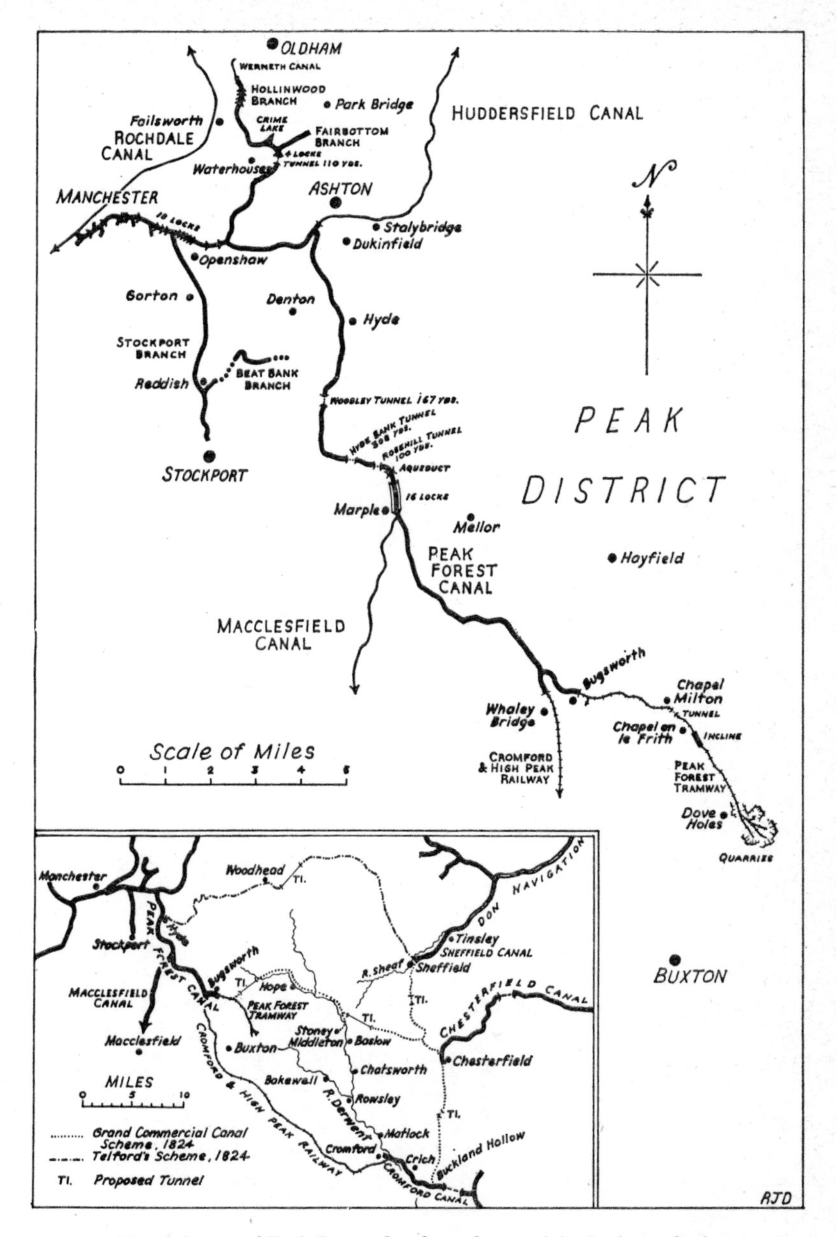

20. The Ashton and Peak Forest Canals, and two of the Peak canal schemes

getting worried because the Rochdale were making no exertions to complete their own section between Piccadilly and Castlefield, and wondered if that company were 'apprized of the forward State of the Ashton Canal Works'. [22]

The Ashton's line was completed to Piccadilly by May 1799, but it was not until May 1800, after a running battle with the Rochdale, that the junction with that canal at Piccadilly was opened, and the Rochdale's line to the Bridgewater available. In December two delegates from the Ashton called on the Huddersfield committee to urge that they should now get to work 'establishing and carrying on a thoroughfare Trade from Hull to Manchester'.[23] They agreed. John Rooth of Manchester also attended, and offered to carry on a 'great trade' from Manchester to Huddersfield if the two companies would help him. From the Ashton he wanted £1,500 spent on building warehouses in Manchester, to be repaid at £150 p.a., and a 10 per cent drawback on account of ground rent he would have to pay. They consented, and Rooth started carrying at the end of September 1801.

In October 1804 the committee met the Huddersfield to discuss making 'such arrangements between the two Companies for increasing the local Trade on the two Canals as may promote the interest of this Canal'.[24] As a result, they agreed to John Rooth's further proposals for carrying on the two canals, and to lend him another £500, and also that the Huddersfield company should have an agent in Manchester. The arrangement with Rooth was discontinued in July 1810, about nine months before Standedge tunnel was completed and through working began.

The canal from Piccadilly to Ashton was 6¾ miles long, the ¼-mile Islington branch coming in at Ancoats. The Stockport branch was 4⅞ miles long, that to Hollinwood 4⅝ miles from the main line at Fairfield, and the Fairbottom, leading off the Hollinwood branch at Waterhouses, 1⅛ miles. This last was not specially authorized, but was built under powers to make collateral cuts. A 3 ft 6 in. gauge colliery tramway ran from it to Park Bridge roller-making works and Rocher colliery. The Beat Bank branch, 3 miles long, was, as we have seen, not finished, though a short section where it came off the Stockport line was later privately owned. Extensive earthworks remain along much of its course.

The main line rose from Manchester by 18 locks, 3 at Ancoats, 4 at Beswick, 9 at Clayton, and 2 at Fairfield, these last being doubled, probably after 1830. The Stockport branch was level, but that to Hollinwood rose by 4 locks at Waterhouses (the centre two being a staircase pair) and another 3 at Hollinwood before the canal ended at Hollinwood basin. Crime lake was later formed there as a result of a burst bank, which caused the valley to become flooded. There were sizeable aqueducts at Ancoats, Beswick, Dukinfield and Waterhouses, and a 110 yd tunnel with a towpath at Waterhouses that was opened out about 1914. There was to have been another, also of about 110 yd, on the Beat Bank branch.

An intake from the Medlock supplied some water to the canal. This was supplemented by lockage water from the Huddersfield and Peak Forest, and by their Hollinwood and Audenshaw reservoirs. They did not, however, get as much from the Werneth colliery as they had expected, and in 1810 bought a beam engine to pump water back up the four locks at Waterhouses to supply the Fairbottom branch and the pound of the Hollinwood line above the locks.[25]

The company's finances took some time to clear up. Because lenders found the security far from attractive, a fourth Act[26] of 1800 allowed £20,000 to be raised by new shares to be issued at any agreed price, ranking pari passu with those existing, and authorized loans to be converted to shares. This Act also empowered the company to charge lime and limestone boats 2s (10p) per lock as well as the tolls. In March 1802 they were so hard up that they could not settle their clerk's legal bill, and had to pay interest on it. Money then trickled slowly in, but it was not until 1805 that the clerk got his yearly account settled for cash. A fifth Act[27] of 1805 stated that £158,655 had been spent out of £170,000 authorized, and that money was still owing. It authorized a new limit of £210,000. The final cost of the canal and its warehouses was probably about £170,000 at opening, and finally rather more.[28] But prospects were better, and in December 1806[29] the price of shares was £95 against an average value of £97 18s (£97·90). A first dividend of £3 per share was paid for that year.

In June 1797 notice was given that on Tuesdays, Thursdays and Saturdays at 08.00 one passenger boat would leave Heaton Norris (Stockport) and another Ashton for Ancoats, both returning at 18.00 hours, while on Sundays there would be a service between Ancoats and Fairfield, Heaton Norris and Fairfield, and Heaton Norris and the junction with the main line at Clayton.[30]

Soon after the canal had been completed from Ancoats to Piccadilly, packet-boat services were revised in July 1799[31] to provide for runs between Piccadilly and Stockport, Hyde Lane on the Peak Forest Canal, and Stalybridge on the Huddersfield, fares being

1s 3d (6p) front room, and 9d back room, from Manchester to Stalybridge or Hyde Lane, 1s (5p) and 8d to Ashton or Stockport. The Peak Forest service (by then extended to Marple) was running in May 1800, but neither it nor the Stalybridge run is mentioned in the following extract from the *Manchester and Salford Directory* of 1802:

'An elegant Boat to convey Passengers and their Luggage, leaves Piccadilly, Manchester, on Sunday and Wednesday mornings, at 8 o'Clock, for Ashton; returns at 4 o'Clock and arrives in Manchester at 6 o'Clock the same evening. Leaves Manchester on Friday Mornings, at 8 o'Clock, for Stockport; returns at 4 o'Clock, and arrives in Manchester at 6 o'Clock the same evening. Passage Boats also leave Ashton and Stockport, for Manchester, on Tuesday, Thursday, and Saturday Mornings, at 8 o'Clock, and return from thence during Winter at 4 o'Clock, and in the Summer at 5 o'Clock in the Evening. The same Boats also leave Ashton and Stockport, for Manchester, on Sunday Mornings at 9 o'Clock, from whence they return at half past 4 o'Clock in the Evening, during Winter and in the Summer at 5 o'Clock'.

	Front Room	Back
Fares	1s to Ashton, 1s 6d return	8d (1s)
	1s to Stockport, 1s 6d return	8d (1s)

The service seems to have been short-lived. It is not mentioned in later directories.

In July 1801 the Peak Forest company had decided to discuss with the Ashton ways of promoting the sale of lime and limestone in Manchester, to the profit of both companies. They were already selling stone and paving there, and soon afterwards got a contract from the Duke of Bridgewater to deliver 400 tons a week for 40 weeks to Castlefield. This seems to have been more than the Peak Forest boats could handle, because at the end of November James Meadows was sent by the Ashton company to ask the Peak Forest to procure enough boats to complete the limestone contracts already made and to provide for the increasing demand. He was to add how necessary it was for that company to build warehouses for the trade from Manchester to the Peak Forest, to finish their reservoir, and to raise the banks of their lower pound to allow its water to run level with that of the Ashton. The Peak Forest company in 1802 bought land in Ancoats Lane for wharves, and the Ashton then decided that 'the Interests of this Canal require that a way and road for Carts and Carriages should be made as soon as possible from the Company's Wharfs near Piccadilly to Ancoats

Lane and that the best line for such way and road will be under the Aqueduct Bridge made over Shooter's Brook'.*[32] They bought more land in 1823.

In January 1803 the Ashton company suggested regular meetings with Huddersfield and Peak Forest representatives. These companies agreed, and thereafter contact was close, especially perhaps with the Peak Forest from March 1808 when the two met 'to arrange a Plan for establishing and encouraging the trade and merchandise' upon the two canals.[33] In November 1805, indeed, the Ashton had agreed with the Peak Forest that James Meadows could also be the latter's agent. Each would pay half his salary of £630 with house and horse.

In December 1805 it was clear that the Werneth Colliery company had been sending coal by road, in spite of their agreement to transport it all by canal. The Ashton in July 1806 therefore decided to end the agreement of 1795 and so the colliery's 6d drawback. This in turn led to less mine water being provided. However, the Werneth company went on using the canal. By 1828 tramroads had also been built from Hollinwood to Chamber and Copster Hill collieries, and from Crime lake to another near Limeside.[34] Later the Werneth concern was absorbed by the Chamber Colliery company, who worked their own boats on the canal until the present century.

In September 1806 the company noted that it would be highly advantageous to join their Hollinwood branch to the Rochdale Canal, and proposed to include authority to do so in a forthcoming Bill, which also sought powers to increase lime tolls. But it was opposed by the Huddersfield and dropped.

By 1807 some Sheffield trade seems to have been going from Manchester via the Ashton and Peak Forest canals, for the company decided to negotiate upon through tolls with the Peak Forest, and in December 1810 the company met promoters of the proposed High Peak Junction Canal.

At the end of 1808 the company were supporting the Leeds & Liverpool's proposed branch from Wigan to Leigh to connect their canal to the Bridgewater. The Ashton evidently saw the chance of some through trade, and sent a delegation to impress on Bradshaw the need for low through tolls. They had a case, for the Bill enabled the Bridgewater trustees to take a compensation toll of 1s 2d (6p) a ton at the junction at Leigh as well as that of 1s 2d also payable at the junction of the Rochdale with the Bridgewater

* Now Store Street.

at Castlefield. Bradshaw would not give way, but orally agreed not to exact more than one charge for goods for Manchester or Liverpool. The Ashton agreed in this case not to oppose, but to get a written promise should the Bill pass. It did not. When the Bill of 1819 was introduced, the Ashton, Huddersfield and Peak Forest all opposed it until it was conceded that only one 1s 2d compensation toll should be payable.

Standedge tunnel, and so the through route by the Huddersfield, being open in 1811, the company initiated talks that year with the Ashton and the owners of Sir John Ramsden's Canal, in order to get through rates down to a level competitive with the Rochdale route. This became a continuous process. In 1825 the Huddersfield asked the Ashton to follow their own practice of not charging for empty boats, to reduce their wharf and labour charges at Stockport —1s 6d a ton against 10d on the Rochdale's wharf at Manchester— to build another warehouse in Manchester as well as the one they themselves were building there, increase the hours their canal was open, and give greater facilities at night and on Sundays. They also observed that the Ashton's locks were very congested, and suggested that they should be doubled; otherwise the Ashton company might find themselves encouraging 'a new communication by a Rail Road or otherwise from the Huddersfield Canal to Manchester'.[35] The Ashton agreed at once to let boats pass on Sundays, to issue night permits when necessary (though they remained sticky about night working), and not to charge empty boats.

In March 1820 Thomas Meadows was brought in to help his father James as agent to both companies. In November 1824 James senior resigned because of ill health. Thomas took over, but died in 1831, when James junior was appointed agent for both canals at £400 p.a., a separate engineer to be employed. This was John Wood, taken on in December at £180 p.a.

In January 1823 both companies decided that the proposed canal from the Bridgewater at Stretford or Sale Moor to Stockport or Hope Green (Poynton) and railway thence to Kerridge near Bollington would be highly prejudicial to them, and agreed to combine in opposition.

The year 1825 brought three schemes, all of which promised additional traffic to the Ashton. The first became the Macclesfield Canal, the second the Cromford & High Peak Railway (see p. 320); but the third, for a canal from the Peak Forest to Sheffield, was stillborn (see p. 319). Then on 9 November 1831 the Macclesfield Canal was opened, and the Ashton became involved in price cutting

Page 303 (*above*) Looking down the Waterhouses locks at Daisy Nook on the Hollinwood branch of the Ashton-under-Lyne Canal in 1928; (*below*) maintenance men repairing a collapse at Hyde Bank tunnel on the Peak Forest Canal in 1895

ALBERT WOOD, PEAK FOREST WHARF, DUCIE STREET, LONDON ROAD, MANCHESTER.

Daily Express Service of Boats

National Telephone: 4934 MANCHESTER.

TO AND FROM

MANCHESTER, ASHTON, HYDE, MARPLE, STRINES NEW MILLS, BUGSWORTH, & WHALEY BRIDGE;

Also BOLLINGTON, MACCLESFIELD, etc., etc.

Goods received at Peak Forest Wharf, Ducie Street London Road, Manchester, up to 6 p.m. (2 p.m. on Saturdays).

PLEASE CONSIGN PER ALBERT WOOD'S BOATS.

Page 304 Albert Wood's advertisement of *c.* 1885. The warehouse on the right was demolished about 1939 between the Ashton–Peak Forest–Macclesfield line and the Trent & Mersey–Bridgewater route to Manchester. From 1839, too, they became involved with other canal companies on the London to Manchester route in trying to work together to quote rates competitive with the railways who were doing the same. In that year they also joined the Peak Forest in asking the Manchester, Bolton & Bury, unsuccessfully, to lengthen some of their locks so that full-length narrow boats in pairs could use them.

At that time the Ashton were getting rather better tolls[36] per mile for London traffic than their neighbours:

	miles	*toll*		
		s	*d*	
Ashton	6⅛	1	1	(5½p)
Peak Forest	8⅛		10½	(4½p)
Macclesfield	26¼	2	2¼	(11p)

In 1828 the company paid 4 per cent, in 1829 5 per cent, and in 1838 7 per cent. Gross revenue in 1828 was £10,736 from 274,020 tons, in 1838 £17,363 from 514,241 tons. But railways as a serious threat were not far away. The Sheffield, Ashton-under-Lyne & Manchester Railway was opened on 17 November 1841 past Guide Bridge, near Ashton, and on 31 March 1842 the Oldham branch of the Manchester & Leeds. Both seriously affected the canal's trade.

After the opening to Guide Bridge, John Boulton of Ashton started a horse-drawn passenger boat service from the station there for 1¼ miles to Ashton using old swift boats from the Glasgow, Paisley & Ardrossan Canal. He built up a good business with the financial help and co-operation of the railway company,[37] till the Guide Bridge to Stalybridge rail branch was opened on 23 December 1845.[38]

In 1842 the dividend had come down to 6 per cent, and p ospects were poor. When, therefore, the Manchester, Sheffield & Lincolnshire Railway (as the Sheffield, Ashton-under-Lyne & Manchester had become) offered to buy the canal for an annuity of £12,364, equivalent to a guaranteed 7 per cent, and another £540 p.a. interest on their debt of £12,000, they accepted in 1846. The purchase was authorized by an Act[39] of 1848, which provided that the canal should be kept open and in good repair, and authorized a new schedule of tolls. Takings for that year fell back to £11,443 and the tonnage carried to 466,188. The shareholders had reason to be satisfied with their bargain. (*To continue the history of the Ashton-under-Lyne Canal, turn to p.* 441.)

Peak Forest Canal

A week before the Huddersfield, the Peak Forest's Act[40] was passed on 28 March 1794. The Ashton's shareholders had begun by planning it as a branch of their own canal, and at their meeting on 8 May 1793 had agreed to introduce a Bill. Afterwards they must have felt that with their other commitments they had taken on too much, for a separate subscribers' meeting was called in July to consider plans and estimates.[41] Ashton shareholders, however, supported the scheme to build a line from that canal's Dukinfield aqueduct by way of Hyde and Marple to Chapel Milton, whence a tramroad was to run to limestone quarries to be opened near Doveholes. Its main purposes were to supply limestone for burning into lime for farming, industrial and building use, and to serve the mines and industries between Hyde and Dukinfield. Two months after the company's first meeting, the enterprising Samuel Oldknow was elected to the committee. Not long before he had acquired extensive property at Mellor and Marple, had built a cotton mill at Mellor, and intended to construct limekilns at Marple to burn limestone brought there by the canal, using his own coal from small local collieries.[42] Oldknow was associated with the Arkwrights of Cromford and, considering how heavily indebted he was to them in 1794, one is surprised that he could support the canal as heavily as he did.[43]

The Act authorized a capital of £90,000, and £60,000 more if necessary. The canal was to run level from its junction with the Ashton through three short tunnels, Rosehill[44] (100 yd),* Woodley (167 yd) and Hyde Bank (308 yd) and then over the great three-arched Marple masonry aqueduct nearly 100 ft above the River Goyt to the foot of the 16 Marple locks with their rise of 209 ft. At the top, 518 ft above sea level, it ran level again to Bugsworth (now Buxworth). By a decision of July 1795 it was decided to end the canal there, and cover the remaining distance to Chapel Milton by tramroad. This canal line was 14¾ miles long as built, to which must be added the short ½ mile branch from near Bugsworth to Whaley Bridge.

> George Borrow was later to write that the greatest aqueduct 'is the stupendous erection near Stockport, which . . . filled my mind when a boy with wonder, constitutes the grand work of England, and yields to nothing in the world of the kind, with the exception of the great canal of China'.[45]

* Rosehill, between Hyde Bank and the Marple aqueduct, was later opened out.

The engineer of the canal until 1801 was Benjamin Outram, who had probably been recommended to Oldknow by the Arkwrights, since he had helped to build the Cromford Canal with which they had been concerned. Thomas Brown was resident. Cutting began from the Ashton junction at Dukinfield, and also upon the Marple–Bugsworth section. By 1795 a lease of land for quarrying had been signed, and before that Outram had been commissioned to supply 'Cast Metal Railways' from the works at Butterley in which he was a partner.[46]

Now appears on the scene an odd but famous character, the American Robert Fulton, to whom has sometimes been undeservedly attributed the invention of the tub-boat canal, the canal inclined plane and the boat-lift. In October 1794, with Charles McNiven (who soon retired from the partnership), he accepted a contract for cutting part of the Peak Forest. In March 1795, however, Outram reported to the committee that 'Messrs. Fulton and McNiven are not proceeding in the cutting that part of the said Canal contracted to be cut by them',[47] which was not perhaps surprising considering McNiven's other engineering commitments on Manchester canals and Fulton's bid in February to cut part of the Gloucester & Berkeley Canal.[48] This was not accepted, but it must have taken time from his work.

Fulton then began to give the committee engineering advice which for a time they liked, but which must have been extremely annoying to Outram. Although the contract for the Marple aqueduct had been awarded on 5 February to William Broadhead and William Anderson, the committee resolved in April

> 'That Mr. Fulton's Idea of substituting Cast Iron instead of Stone Arches for the Aqueduct over the River Mersey* is approved, and that the Engineer and Mr. Fulton be desired to give the Clerks instructions for the particulars of an Advertisement for procuring Offers of Terms for furnishing Cast Iron for making the same'.[49]

He then probably suggested the possibility of using tub-boats and an inclined plane, for Outram was sent with two others 'to view the operation of Small Boats and inclined planes at Coalbrook Dale and to report the result'.[50] They returned to say that small boats and inclines would be practical, and would save money. The committee postponed its decision, but 'Resolved that Mr. Fulton be requested to print his Ideas of the comparative Advantages of

* The River Mersey is now taken to end at Stockport, where it divides into the Goyt and the Tame, the aqueduct being over the Goyt.

twenty Ton Boats and five Ton Boats with two Ton Boats and Railways and that Mr. Fulton do get 200 Copies thereof printed at the Expense of the said Company'.[51] This idea of writing down his ideas probably became first the article on 'Small Canals' he wrote for the London *Star* on 30 July 1795 and then the book, *A Treatise on the improvement of Canal Navigation*, which he published in 1796, and which may have been paid for by the Peak Forest Company.

In June[52] the company voted him 100 guineas for 'having suggested to this Committee . . . many Ideas in and about the execution of the said Canal and its works and drawn and produced to this Committee many plans', and in September he was mentioned as still working as a contractor.[53] He then disappears from the Peak Forest records, probably to Outram's relief. The aqueduct was not built with iron arches, and inclined planes and tub-boats were not employed on the canal, though a few tub-boats may have been built.[54]

During the first half of 1796 a proposed Macclesfield Canal to link the Peak Forest at Marple with the Caldon branch of the Trent & Mersey caused a flurry of activity,[55] but this died quickly away as money became tighter. By July the Peak Forest itself was short of cash, and members of the committee had to lend it money to complete the summit level and the tramroad, which were opened from 'the lime works' via Bugsworth to Marple on 31 August 1796.[56] Samuel Oldknow started to work his limekilns about this time. Coal and limestone were carried to the kilns on two arms leading out of Marple basin, while lime and lime ash were loaded on the short arm running from Possett bridge. About half the kilns' output was used for agricultural and other local purposes, the rest being sent away by boat. In 1811 Oldknow leased the kilns to Wright & Brown (the latter being the canal's resident engineer), who managed them until about 1860.[57]

In March 1797 work had to be temporarily stopped on the lower level for lack of money. At this time about £80,045 had been raised from the shareholders in 806 shares of £100 each, less nine that had been forfeited. It was agreed to issue twice as many new shares as the 797 in existence, or 2,391 in all, to be allocated to shareholders willing to pay up to £50 each for them. Money was tight, and Oldknow was not the only shareholder behind with his calls. It was therefore decided to finish the lower pound and the aqueduct, but to build a single-track tramroad to connect the upper and lower pounds at Marple instead of 16 locks. In January 1798 Outram's estimate for this tramroad was accepted, and in April toll reductions

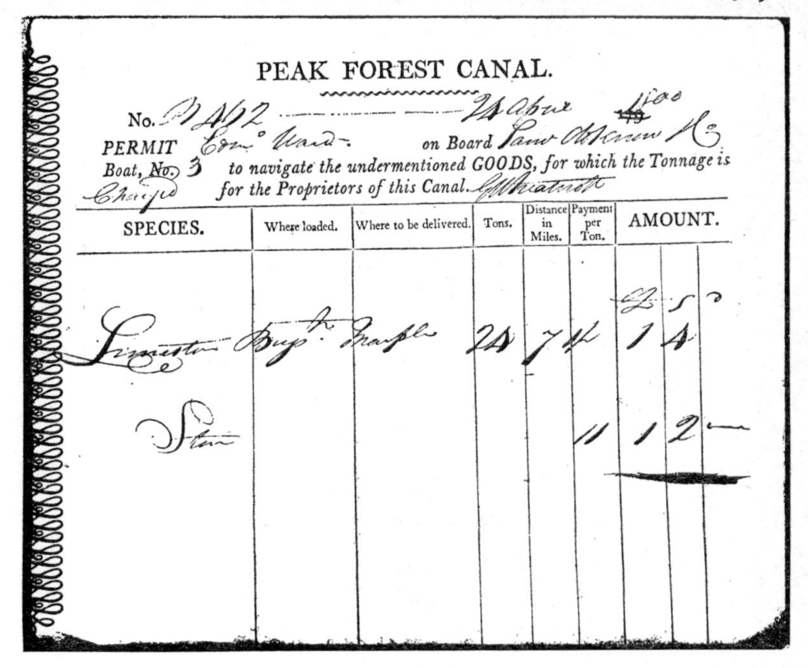

PEAK FOREST CANAL.

No.

PERMIT on Board

Boat, No. 3 to navigate the undermentioned GOODS, for which the Tonnage is

for the Proprietors of this Canal.

SPECIES.	Where loaded.	Where to be delivered.	Tons.	Distance in Miles.	Payment per Ton.	AMOUNT.

21. A permit of 1800 for 24 tons of limestone sent from Bugsworth to Marple on one of Samuel Oldknow's boats

were made on lime, limestone, coal for burning in kilns, and building stone, to help trade develop.

The lower level except for the Marple aqueduct was in use in 1799. On 1 May 1800[58] the whole canal, including the Whaley Bridge branch, was opened with its intervening tramroad. A month later an Act[59] noted that £117,140 had been subscribed, £80,600 from calls and the balance in loans, and gave the company powers to raise their original total of £150,000 either by new shares or promissory notes.

The main quarry tramroad[60] was 6½ miles long from Bugsworth with its basin (two more were added later) for loading limestone and lime, warehouse, sheds and kilns,[61] by Chapel-en-le-Frith to the quarries at Doveholes, with branches to other quarries. It reached a summit level of 1,139 ft. There was an 85 yd tunnel at Chapel Milton and a 209 ft inclined plane at Chapel-en-le-Frith. The track was of 4 ft 2 in. gauge, a plateway of the usual Outram type. It was doubled in 1803, except for the tunnel.

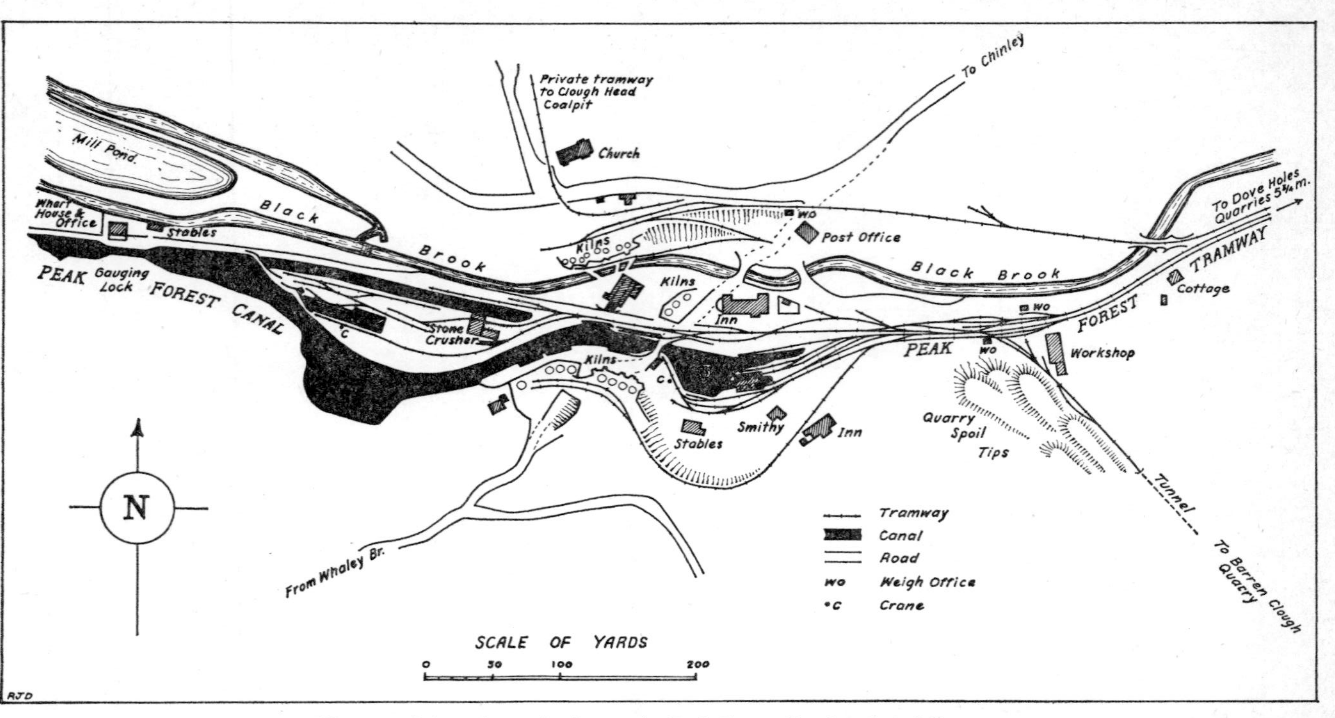

22. Bugsworth interchange basins on the Peak Forest Canal at their fullest extent

The working of the tramroad by the canal company has many points of interest, though what is recorded as happening in the early years did not necessarily go on happening. The quarries were on land owned or leased by the company, and the quarrymen were employed by them on group contracts. By October 1800 their number was to be increased to 200, but it cannot have been easy to get labour. An advertisement of November 1800 seeks quarrymen at very handsome wages, and adds:

> 'As a further encouragement, a quart of good, wholesome ale per day will be allowed to each workman, over and above his wages; and a Flannel Waistcoat and Trousers will be given to such Workmen as may contract for working by the piece.'[62]

Much trouble was also taken over their conditions. Some of the houses in which they lived were rented for them by the company, and others were furnished with beds. A store was also opened at which they were to be supplied with goods at cost price.

The limestone was carried down the tramroad in iron boxes carried in waggons taking 2–2½ tons in all. At Bugsworth the containers were either lifted into boats or taken straight to kilns. The ganging of the stone down the tramroad was, at any rate at first, done by contractors, and limestone was sold at Bugsworth at prices which could include tolls if required. Limestone loaded on to the canal might be going to Oldknow's kilns at Marple, or down the Marple tramroad in boxes to be loaded again into boats at the bottom. Boxes were not compulsory, but tolls were lower if they were used. Their use probably ended after the Marple tramroad had been closed, for transhipment was then no longer necessary, waggon tipplers being installed at Bugsworth instead.

Activity in the limestone and lime trades meant that it became more and more inconvenient to tranship all loads twice, especially since the Marple tramroad was becoming congested with traffic (the company wanted at least 50 waggons kept at work on it, and a minimum of 1,000 tons of limestone a day handled), and night working had to be introduced in December 1800. In 1801, therefore, the company not only doubled the Marple tramroad, but started to move for a Bill to raise money to build locks. They were, however, threatened with the opposition of millowners who presumably feared a loss of water, and had therefore to decide to do what they could out of their own resources. Soon afterwards Oldknow and Arkwright jointly were asked and agreed to put up the money and build the locks under the supervision of the company's engineer in exchange for preferential tolls till the money was repaid. Later Oldknow withdrew, probably because his finances were not strong enough,[63] and a final agreement was made with Arkwright in August 1803 to advance £24,000. Thomas Brown was engineer on this work, for Outram had left, and the locks[64] were opened in October 1804[65] at a cost of about £27,000. In February 1807 the Marple tramroad ceased to be used. Meanwhile an Act had been passed in 1805[66] to authorize the raising of more money. It appears that the total cost of the canal including the locks was about £177,000.

In 1811, seventeen years after the company had first been promoted, a maiden dividend of £2 per share* was paid. Two years later Arkwright was repaid.

The canal company went in for interesting sales methods to popularize the use of limestone, and the burning and use of lime. In their early years they offered drawbacks on limestone and coal for those who would build limekilns on the Peak Forest or communicating canals, and soon afterwards to pay part of the cost of new limekilns built within a time-limit, if Peak Forest limestone only were used in them. They offered premiums for boats on the upper level built before a certain date and loans for building them and bought some for their own trading. They requested Outram to ask Dr Anderson 'to write and publish a Treatise (which may be afforded to be sold for One shilling) on the nature effects and proper Application of Lime as a Manure', or to allow extracts to be made and published from his existing works, offering to see that 1,000 copies were sold. Later, they made agreements with other companies for sales at special rates that included the cost of stone, tolls and freight by which Peak Forest limestone was sold as far afield as Preston Brook and Runcorn on the Bridgewater, the Leeds & Liverpool beyond Leigh, beyond Standedge tunnel on the Huddersfield, and the summit levels of the Rochdale and the Manchester, Bolton & Bury canals. They bought additional estates for quarrying; for a time they rented kilns to operate themselves in addition to those they already owned at Bugsworth. Lastly, they maintained their own wharf in Manchester on land near Ancoats bought from the Ashton company, where limestone was sold for road-building. For this purpose they made bulk delivery contracts at cheap rates with turnpike trustees to enable them to experiment with limestone for roads. Later, when the Liverpool & Manchester Railway opened, they worked with it, granting drawbacks on

* The average value of a share was £78.

PEAK FOREST CANAL.

Feby 6th 1809

No.
PERMIT
Boat, No. [illegible] on Board Saml Oldknow's
to navigate the undermentioned Goods, for which the Tonnage is
Charged for the Proprietors of this Canal. Wm Hamilton

SPECIES.	Where loaded.	Where to be delivered.	Tons.	Distance in Miles.	Payment per Ton.	AMOUNT.		
Limestone	Bugsworth	Marple	23	7	7	–	13	5
Good Stone Weight R Goodwin		Limestone			2/2	2	9	10
					£	3	3	3

the Honesty of Marple

23. A permit of 1809 for a cargo of limestone for Samuel Oldknow, boated in the *Honesty of Marple*

limestone carried by canal to Manchester and then put on rail for Liverpool, or the Warrington or Bolton lines.

The limestone and lime trade was a busy one. In 1808, 50,000 tons of limestone came down the tramroad; in 1824 the company loaded 291 narrow boats in one period of four weeks, and in 1833 was carrying an average of 1,743 tons, and loading 279 boats, weekly.[67]

In 1823, when the canal was more prosperous (the dividend for 1822 was £3 10s—£3·50), two developments came nearer to fruition. The Macclesfield Canal to link the Peak Forest to the Trent & Mersey, which had been talked about for thirty years, now began to be seriously pressed, and the committee recorded that they 'view the said Undertaking as holding out prospects highly beneficial to this concern'.[68] It was authorized in 1826 and opened in 1831. They welcomed the Cromford & High Peak Railway also 'which if made and completed would be of great benefit to this Canal'.[69] This railway through the Peak District to link the Cromford and Peak Forest canals (see p. 320) was seen at the time as helping trade between Manchester and the Nottingham–Derby area, and also as part of possible London–Manchester and Manchester–Sheffield through canal and rail routes. The opening of the Macclesfield Canal involved the Peak Forest, as its ally, in a rate-cutting war with the Trent & Mersey, operating the older canal line from the Potteries to Manchester. Again, after the Cromford & High Peak Railway had been opened in 1831, that company and the Peak Forest found themselves allied against the Macclesfield and the Trent & Mersey in competing for the trade to the Nottingham area in textiles and Staffordshire iron, and from it to Manchester in flour and malt. Nevertheless, coal, lime and limestone continued to be the canal's main traffics. Their growing volume, though it put pressure on water supplies, enabled the dividend to be raised from the £2 of 1827 to £3 10s (£3·50) for 1832, £4 for 1833 and 1834, and £5, the highest reached, for 1835 to 1837. Tonnage carried in 1838 was 442,253½, and the total revenue £19,169.

The Peak Forest had also become, since the opening of the Macclesfield in 1831, part of a through canal line from Manchester to London in competition with another by way of Preston Brook and the Trent & Mersey. The opening of the Grand Junction Railway from the Liverpool & Manchester line to Birmingham in 1837, and the completion of the London & Birmingham Railway in 1838, produced a through rail line that was ready to compete with both these canal routes, three railway against eight canal companies on the route via the Peak Forest. The Grand Junction Railway began the rate-cutting, and the Peak Forest, with more initiative than the Ashton, and less tied to the policy of the Trent & Mersey than the Macclesfield, took the lead in organizing canal resistance. Some of their efforts to arrange common toll reductions and create a common policy have been described in *The Canals of the West Midlands*.[70]

Meanwhile heavy competition was developing on the Manchester and Sheffield route via the Trent & Mersey, the merchandise toll being reduced to 1d in April 1841 and to ½d in October. It developed also on the Manchester–Cromford and Manchester–Nottingham route between the Cromford & High Peak Railway on the one side and the canal line via the Trent & Mersey on the other. The Peak Forest was on the side of the High Peak Railway here, and reduced its toll to 1s (5p) throughout and 3d wharfage for interchange at Whaley for goods from Manchester to Cromford and beyond.[71] Now the lime trade also began to be affected by railway competition, first to Rochdale from the Manchester & Leeds Railway, then elsewhere, and the canal company therefore increased the drawbacks it offered. In April 1843 'close competition with Railways and other lines of Communication'[72] was reported.

Late in 1843 and early in 1844 the Peak Forest and Ashton companies considered jointly leasing the Cromford & High Peak Railway, which was in low water. However, there was a threat of a railway from the Manchester & Birmingham line near Poynton to Buxton, and they decided against. By 1844 toll reductions had so increased traffic on the canal that receipts equalled those of the previous year, much to the shareholders' relief. However, dividends were down to £3 a share for 1843 and 1844. So, in the railway mania year of 1845, their minds were receptive to an offer from the Sheffield, Ashton-under-Lyne & Manchester Railway to take over the canal and its debts on a perpetual lease. The annuity offered was £9,325 p.a., which represented a guaranteed dividend of £3 18s (£3·90) per share of an average value of £78, or exactly 5 per cent on the 2,391 shares, plus £1,856 interest on the debt of £41,000. The purchase was completed on 25 March 1846, and the canal was transferred on 27 July. For its last independent year to 25 March 1846, total receipts were £31,216, including those from the stone trade; expenditure was £17,760, £11,344 of which was attributable to the stone trade and £6,416 to the canal.[73]

The same railway also leased the Ashton and Macclesfield canals, and therefore safeguarded its local and Sheffield trade, as well as gaining a still profitable limestone business that was carried on for many years. (*To continue the history of the Peak Forest Canal, turn to p.* 444.)

Compstall Navigation

From about 1820 cotton mills were developed at Compstall bridge on the River Etherow, just above its junction with the Goyt at Marple. They were powered by water stored in two reservoirs immediately above them, these being supplied through a channel from the upper end of the northern reservoir along the river bank to a weir about 1,200 yd above the mills. For a period in the mid-nineteenth century small boats brought coal to the village near the mills from pits near the weir.[74]

Canal Projects in the Peak District
The Cromford & High Peak Railway

As early as 1789, the date of the Act for the Cromford Canal, it was thought desirable to extend it by Matlock and Buxton to Manchester. But this was premature. By May 1800, however, four waterways had been completed to the edge of the highlands of Derbyshire, the Peak Forest to Bugsworth and Whaley Bridge, the Don Navigation[75] to Tinsley between Rotherham and Sheffield, the Chesterfield Canal[76] to Chesterfield, and the Cromford[76] to Cromford. To join the Peak Forest to the Cromford would much shorten the line from Manchester to Nottingham and, indeed, London; to link the Peak Forest and the Don would do the same to the long route via Preston Brook, or the so far uncompleted alternative via the Huddersfield Canal. To bring the Chesterfield Canal farther into the Peak District would be to provide a cheap supply of coal for lime-burning and local use, and from the central district limestone could be carried downwards for burning elsewhere. Lastly, there was plenty of water.

It began with a proposal for a canal from Bakewell to join the Cromford, made in October 1802 at a meeting at Matlock with Richard Arkwright in the chair. The Cromford company, clearly seeing the project as the first step to a junction, agreed to reduce their tolls for the new canal's traffic, and to subscribe £10,000 out of an estimated £60,000 needed.[77] A few days later the Peak Forest agreed to put up £15,000.[78] In May 1803, however, the Bakewell supporters decided not to proceed. The Cromford acquiesced, but resolved to offer toll concessions should it be revived.

The idea was revived in 1810 as the High Peak Junction Canal by the Grand Junction Canal company when they were supporting a Bill to authorize the Grand Union, last link in the Leicester line of canals that would give them a direct link with the Trent & Mersey, and also with the Erewash and Cromford canals.[79] They envisaged the Cromford being continued to the Peak Forest and so to Manchester, giving an alternative route for London trade to that by the Oxford, Coventry, Trent & Mersey and Bridgewater canals. It was a tactless moment to choose, for the Trent & Mersey at once opposed the Grand Union Bill, and only desisted when the Grand Union and Grand Junction promised them and the Bridgewater trustees not to support the High Peak Junction.[80]

During the summer, however, the idea had gained its own momentum. In June, a month after the Grand Union's Act had been passed, it was considered by the committee of the Peak Forest, who suggested that the London supporters of the project should seek the backing of all canal companies likely to be interested. The committee also instructed the Peak Forest's engineer, James Meadows, to survey two possible lines, one from Bugsworth by Edale and the Derwent valley, the other by Buxton and the Wye valley. The project was clearly still thought of as a link in a London canal line, but the committee suggested that a branch should be made to Chesterfield.[81] The Cromford company were now interested, but not active in support. By the autumn, Meadows and Nicholas Brown had done surveys, and the line had also been re-surveyed by Rennie.[82]

Rennie's line was from Chapel Milton (whence the Peak Forest Canal would be extended under the powers of its existing Act) through a 2¾ mile tunnel to Edale, then down the Hope and Derwent valleys. It crossed the Derwent near Baslow, and ran through a 1 mile tunnel to the Wye valley near Bakewell (this to avoid Chatsworth) and down to Great Rowsley to re-cross the Derwent and follow it to Matlock. One more tunnel, 1½ miles long, then took the line to the Cromford Canal near its Derwent aqueduct. Its length was 38¼ miles, with a rise of 149 ft from the extended Peak Forest, and fall of 530 ft to the Cromford Canal. His estimate for a narrow-boat canal was £650,000, less if the line could go through Chatsworth park across the river from the house.[83] There would also have been the cost of extending the Peak Forest Canal to Chapel Milton. A parliamentary notice was issued in

November, and a meeting of the London and Manchester committees called for 19 December at Derby,[84] but by then the scheme was dead. In 1813, however, when an enquiry about it was made to the Peak Forest, the committee said they believed the promoters 'have not abandoned the Idea of carrying that scheme into effect'.[85] Indeed, a month later William Chapman the engineer produced a report which suggested linking Sheffield with Manchester by a railway from that town up the Sheaf vale, rising by six inclined planes worked by steam engines, and then passing through a 2¾ mile canal tunnel under the East Moor ridge and into a short canal to join the High Peak Junction.[86] The traffic foreseen was apparently limestone.

In the autumn of 1810, a little before the collapse of the High Peak Junction scheme, there was a proposal for a canal, the North Eastern Junction, to run from the Cromford Canal at Pinxton to the Chesterfield Canal and, leaving that again at Hillamarsh, to join the River Don at Rotherham,[87] so giving access to the Dearne & Dove Canal. The first public meeting was held at Rotherham on 26 October 1810, with Richard Gresley in the chair, and proposed wide tunnels so that both wide and narrow boats could use the canal. The meeting was thinly attended, and the Derby paper, reporting the meeting, cautioned its readers against 'the speculative suggestions of perfect strangers'.[88]

Rennie surveyed the line, and reported that he had found a favourable route.[89] A few months later the Cromford company were considering the arguments in its favour:

> 'the Yorkshire Trade which now chiefly goes by the uncertain Route of the Humber to London, and is generally from ten Days to a Fortnight . . . would . . . arrive in London with Certainty . . . in the space of one Week. The Trade from Sheffield to Birmingham, instead of going as it now does, chiefly by Land Carriage, in four or five Days, at an Expence of five shillings per Hundred Weight, would go with perfect Safety in three Days, and at less than one Half of the Expence'.[90]

They thought the idea beneficial and sought the Duke of Devonshire's support. In June 1811 the Nottingham Canal company recommended their shareholders to 'give the Measure every reasonable support'.[91] On the other hand, the Don Navigation company, who had been approached by Gresley in November 1810, showed no perceptible interest. In the form of a possible junction between the proposed Sheffield Canal* and the Cromford

* From the Don Navigation near Tinsley to Sheffield, authorized in 1815.

Canal, the scheme was still being discussed in September 1814.[92]

In the early twenties the Chesterfield Canal company were considering an extension 'to the Barlow, and its neighbouring, Coal Field . . . and from thence to the Calver or Middleton Limestone Rocks' and the Sheffield Canal company had various extensions to the west surveyed. These ideas were caught up with the older ones in the spring of 1824, in a proposal by Joseph Haslehurst, engineer of Unstone colliery, Chesterfield, which was set out in a prospectus of 24 June, for the Grand Commercial (or Scarsdale & High Peak) Canal, 44 miles long. It was to run from Bugsworth on the Peak Forest Canal via the Hope and Derwent valleys to Grindleford, and thence to the head of the Barlow brook, whence one line would join the Sheffield Canal, and one go to Chesterfield and on to the Cromford Canal at Buckland Hollow. The distances by this narrow-boat canal from Bugsworth would be 33½ miles to Sheffield, 28½ miles to Chesterfield and 44¾ miles to Buckland Hollow, with nearly 10 miles of tunnelling. The estimate was £574,130, the estimated annual income from limestone, coal, stone and merchandise being £62,250 p.a.[93]

Later in 1824, Haslehurst elaborated his proposal and slightly altered the figures.[94] Traffic he saw as mainly lime and limestone, coal and merchandise, with corn, malt and flour especially on the main line.

He pointed out that his scheme reduced the waterway distance from Manchester to Stockwith (where the Chesterfield Canal joined the Trent near Gainsborough) from 206 miles to 95½, that to Sawley (where the Leicester line joined the Trent) from 124 miles to 86¼ miles, and to London by 6 miles.*

A meeting was held at Buxton on 15 July which resolved that this and various other plans and schemes should all be submitted to Telford for his report and suggestions.[95] A meeting at Sheffield in August raised some money towards Telford's fees.[96] One in Manchester at the end of October, with Samuel Oldknow in the chair, contributed some more.[97] On 12 November another meeting in Sheffield considered a canal or railroad—it must benefit Sheffield, as well as join the Peak Forest, Chesterfield and Cromford canals—and the newspaper commented 'Everything now depends on Telford's Report'.[98] It was followed by one on 1 December at Manchester which decided that a railroad was the answer to a communication between the Peak Forest and Cromford canals, and subscribed £150,000.[99]

* The Macclesfield Canal had not then been built.

So it was that by the time a meeting at Buxton in February 1825 heard Telford report upon a canal line from Sheffield via Penistone to the Peak Forest Canal at Hyde, the railroad scheme launched in June 1824 that became the Cromford & High Peak Railway had overtaken it.

This, though not the active concern of either the Peak Forest or the Cromford companies, was to run from the former's Whaley Bridge terminus to the Cromford at High Peak junction near the Derwent aqueduct. It was thought of as a locally useful line, and also part of a through canal/rail route from London and Nottingham to Manchester. Its potential usefulness was also apparent to railway promoters.

The Cromford & High Peak Railway was authorized in 1825,[100] engineered by Josias Jessop, and opened on 6 July 1831 with nine inclined planes. It was worked by horses until 1841, and then by locomotives. The cost of construction was £159,327 to August 1831, the revenue nowhere near the £16,676 p.a. of the estimate. Apart from local trade, the only through carriage that seems consistently to have been moved in early days was in flour and malt from the Nottingham area to Manchester. By 1832, when the opening of the Macclesfield Canal had increased competition, German Wheatcroft,[101] carrying on the Nottingham Canal, asked that his boats might have rate concessions and pass at all hours 'to enable him to promote a new trade on the High Peak Railway from the Peak Forest Canal, and to compete with traders on other canals who are permitted to pass at all hours'.[102] He got some concessions. In 1841 the railway company, with the Cromford, Peak Forest and Ashton Canal companies, made considerable toll reductions on flour, corn and malt to Manchester, and on merchandise, and succeeded in attracting traffic from the Trent & Mersey's route.

These efforts did not much help. In 1844 the Cromford & High Peak company approached the Peak Forest offering a lease of the railway with an option to buy, at a time when it was many years in arrears with its mortgage interest and owed £50,000.[103] The Peak Forest nearly agreed, but were frightened off by the promotion of a railway from near Poynton to Buxton. The High Peak Railway was then approached by the Cromford company, who in turn might have accepted a lease if they had not begun to negotiate for their own purchase by a railway company. In 1845 the Peak Forest Canal passed under railway control, and from that date the Cromford & High Peak line ceased to be much of a factor in canal competition with other lines or with railways, though the Cromford company negotiated further reductions on the whole length of the railway as on its own line in corn, flour and malt in 1850. In 1852 the Cromford also was taken over by a railway, and in 1861 the High Peak concern was itself leased by the London & North Western Railway.

The Cromford & High Peak scheme was of no benefit to Sheffield. The Peak Forest company had welcomed Telford's proposed line in November 1825;[104] in April 1826 Henry Sanderson had issued his *Considerations on the proposed communication by a navigable canal between the town of Sheffield and the Peak Forest Canal; with remarks and calculations tending to prove the superiority of an edge railway*. In 1827 a committee was appointed again to consider how the Peak Forest and Sheffield canals might be united,[105] and in 1828 the Sheffield company asked the Peak Forest to share with them the expenses of a survey for a railroad to join the latter's tramroad near Chapel Milton; the Peak Forest declined.[106] One can see in this, as in Sanderson's scheme, adumbrations of the Sheffield & Manchester Railway plan of 1841, with its six inclined planes through the High Peak, and later of the Sheffield, Ashton-under-Lyne and Manchester Railway, opened in December 1845.

CHAPTER XII

The Huddersfield Canal

In the 1770s Sir John Ramsden had constructed his broad canal for $3\frac{3}{4}$ miles from the Calder & Hebble at Cooper Bridge to Huddersfield,[1] and in 1792 Bills for Pennine canals from the Rochdale promoters and a group connected with the Manchester, Bolton & Bury company had both failed. A canal line from Manchester to Hull was badly needed. The Ashton Canal had been authorized in 1792, and a line from it to Sir John Ramsden's would be at least 15 miles shorter than either the Rochdale or the Manchester, Bolton & Bury schemes. Assuming that a connection were made later between the Ashton and the Bridgewater, there would be a through route from Runcorn. The advantages were the short distance and the cheaper cost of building a narrow-boat canal; the dangers, the unknown problems of the necessary summit tunnel, and the possibility of either the Rochdale or the Manchester, Bolton & Bury lines being authorized. If they were, their routes were easier, they had direct access to Manchester and to the Calder & Hebble without an intervening canal at either end, and they could be built to barge width with consequent economies in carrying. If the Ashton shareholders had not wanted to get more business for their line, and if the canal mania and the premium ruling on Ashton shares had not encouraged the promoters, it is doubtful whether the Huddersfield Canal would ever have got started.

But it did. A group, many of them Ashton shareholders, came together in May 1793, for whom Nicholas Brown did a survey and Benjamin Outram a report on 22 October. He estimated the canal at £178,748 exclusive of parliamentary expenses. Knowing that millowners on the line would oppose any interference with their water supplies, he allowed for reservoirs to provide a hundred locksful a day without drawing on millowners' water. Of the necessary summit tunnel near Marsden, he wrote:

'The hill through which the tunnel is proposed to be made

> appears favourable; the strata consist of gritstone and strong shale, and the low ground, in the centre, near Red Brook, will afford an opportunity of opening the works by means of steam engines, so as greatly to facilitate the completion of the tunnel, which, I conceive, may be accomplished in five years.'[2]

On the following day a meeting of subscribers at the George Inn, Huddersfield, resolved:

> 'That from the Plans produced to this Meeting . . . and from the Report of Mr. Outram the Engineer, it is the opinion of this Meeting, that the proposed Canal may be executed in the Line so laid down, without Injury to Mill Property.'[3]

Subsequently, £4,000 was added to the estimate to make the reservoirs larger. Huddersfield shares quickly rose to £15 premium, then fell back. In the meantime, goods were leaving Manchester each morning by road, reaching Huddersfield in the evening for shipment next morning by water to Wakefield and Hull. A similar service operated in the other direction. Rates were 1s 3d (6p) per cwt from Manchester to Huddersfield, 1s 6d (7½p) to Wakefield, 2s 3d (11p) to Hull.[4] The differential must have encouraged the canal promoters.

The Act[5] was passed on 4 April 1794. It authorized a share capital of £184,000, and £90,000 more if necessary. Careful rules were made on water supply, notably by compelling the company to build reservoirs to hold 20,000 locks of water. None was to be taken from rivers on the line except in flood-time. Sir John Ramsden was not to take tolls on the short section of his canal between his wharves and warehouses and the junction with the Huddersfield Canal, but if he failed to maintain it, the latter company could do so. He, the Calder & Hebble and the Aire & Calder were protected against the building of another canal to the east of the Huddersfield line.

This canal, taking 70 ft × 7 ft narrow boats, was to be 19⅞ miles long, rising 338 ft from an end-on junction with the Ashton's main line through short tunnels at Stalybridge* (198 yd) and Scout near Mossley (220 yd) and 32 locks to Diggle and the great summit tunnel at Standedge, 3 miles 176 yd long† and 648 ft above sea level. At Marsden it began a descent of 493 ft by 42 locks to join

* Later opened out.

† In 1893–4 the double-track railway tunnel was built, with a bend at each end which took it over the canal line. To accommodate this at the Diggle end about 32 yd of the old tunnel were removed, and its length was then extended by a 'covered way', as it was then described, built up with steel girders and floor plates, for some 274 yd. The present scaled length is therefore 3 miles 418 yd.[6]

Sir John Ramsden's Canal at Huddersfield, where transhipment would be necessary to and from craft 57 ft 6 in × 14 ft 2 in. At the first meetings Benjamin Outram was appointed engineer on a daily basis, with Nicholas Brown as his surveyor and superintendent. He was instructed to start work at each end of the line, and on 1,000 yd at the west end of the tunnel, which had been contracted to Thomas Lee.

The committee were accustomed to meet at Ashton, Delph, Marsden or Huddersfield. At first things went well enough, but very soon the great tunnel began to overshadow their meetings, and in July 1795 John Evans was given charge of it under Outram at 2 guineas a week—he later did some contracting, and left in 1798. Seemingly it was Outram's original intention to tunnel not only from each end, but from the foot of vertical shafts sunk from the top of the hills, using pumps to get rid of the water he knew he would encounter. The midsummer report of 1796 describes preparations for building nine waterwheels to raise water or spoil from the tunnel, 1,485 yd of small subsidiary tunnels having been driven to supply them and drain surface springs. Fourteen pits were then finished or making, and one large and three small steam engines installed. Only 795 yd had then been cut, and some of it arched, yet £92,000 had already been called and £69,348 received.

By the autumn of that year it was clear how difficult the job was going to be. Much money was being spent 'in endeavouring to raise Water from and to sink some of the Pitts . . . without advancing with the Works in proportion equivalent to the Expences',[7] while in November one pit was being sunk at only a yard a week. It must have been soon afterwards that Outram decided only to tunnel from the ends, using the pits and header tunnels for pumping the great quantities of water he was encountering, or to raise spoil. The decision saved money in sinking pits, but was to add greatly to the waiting time the company would have to endure before the tunnel would be finished.

It seems that the canal was open from Ashton junction to Stalybridge by November 1796, probably soon afterwards from Huddersfield to Slaithwaite, and by August 1797 from Stalybridge to Uppermill near Saddleworth. Wharves were ordered to be built at several points on each side. By April 1797 work was being slowed down as money trickled in, and the treasurers were pressing for a reduction in the company's overdraft. By June £80 had been called on each £100 share, there was only £3,464 in the bank, and arrears of £29,099.

In September Outram told the committee that Thomas Lee, the first tunnel contractor, could not complete his work, 'having been impeded . . . by many unforseen Circumstances'. He had lost heavily and 'will be ruined if some Allowance and Indulgence is not given him'.[8] The committee gave him some money for timber, increased his rate per yard for the rest of his contract, and allowed him an extra year. They also set up a tunnel sub-committee to watch the work. In early 1798 a delegation was sent to the Ashton company to complain that their canal had not yet joined the Rochdale; later that year the Huddersfield–Marsden section was completed, and early in 1799 that from Stalybridge to Woolroad (Saddleworth). Nicholas Brown's job now came to an end. Once working began, millowners started to complain that their water was being taken, and the engineers were told to gauge all supplies from reservoirs to the canal that passed by way of rivers used by mills.

By now the full £100 had been called on each share, but much was owing from shareholders, some of whom were bankrupt, others dead, while a few had left the country. The committee therefore tried to raise £20,000 by mortgage, to be repaid before a dividend was declared, but received only £8,000. They then agreed that each would take a district, and 'use their utmost endeavours by personally waiting on Gentlemen'.[9] They managed to bring the total to £14,182 by July 1798, when they gave notice to forfeit all shares which had paid £60 but were in arrears for the rest, and soon afterwards all that had not paid the full £100. By May 1799, 43 shares had been forfeited. Failure to pay calls led in 1796 to an important court ruling. John Buckley, an original subscriber to the company, subsequently sold five shares to John Kelsall when only £4 had been paid. Kelsall, owing £36 on each for calls, later became insolvent. The company brought an action against Buckley as the original subscriber, but lost it, the court ruling that the current owner of a share alone was liable for calls.

By mid-1799 1,000 yd of tunnel had been completed, and another 1,000 yd opened; all headings had been driven, and all pits completed except parts of three. In January two trading boats had been ordered by the company to get business started on the Huddersfield side, and in February they arranged to begin regular through working, agreeing with William Davenport to carry 'with all possible Expedition and care' all goods he was given between Woolroad and Marsden at 6s 6d a ton. The service was advertised to start on 23 April.[10] Tolls were cut, and sub-committees asked to do all they could to encourage both local and through traffic. In August the committee were trying to get a daily boat running between Manchester and Woolroad. Then came 'unprecedented' floods. The works were seriously damaged and part of the aqueduct between Ashton junction and Stalybridge was carried away. Not enough money was in hand to pay for repairs and re-establish a trade; it would be necessary to go to Parliament for additional powers. In the meantime the committee asked shareholders for a voluntary 5 per cent call, to finance some £6,500 worth of repairs and precautionary works against another such disaster; these included a cast-iron arch, recommended by Outram, for the aqueduct. The resulting Act[11] of 1800 did not increase the company's original authority to raise £274,000, but enabled them to obtain money by extra calls on shares up to £20, or by new shares or promissory notes. These calls were made, giving the 1,691 remaining original shares (149 had been forfeited) a value of £120 each.

In early 1800 money was so short that the men engaged in repairing the flood damage could only be paid monthly. On one occasion work had to be stopped, but in May it restarted, and in June a call could be made under the new Act. The finished sections were not fully reopened, however, until early 1801, and so hard up was the company that in December they could only pay £30 each towards their debts to the workmen, bigger debts receiving 5s (25p) in the £ on the balances. The men were not finally paid off until March 1801. In June 1800 we find a name as a small contractor for flood damage repairs which was later to be famous, Edward (later Sir Edward) Banks.

Worried by their troubles, the committee asked Outram to be 'upon the line of the Canal as soon as possible, and that he will give all the time and attention in his power to the Works of the Canal'.[12] They asked him to continue excavating the tunnel at both ends at a cost that they hoped would soon be covered by the profits earned from the parts already open. Meanwhile, having themselves built a tramroad at Marple to connect two levels of their canal, the Peak Forest company, who had a trading interest in the completion of the Huddersfield, in October 1800 suggested a tramroad from Woolroad to Marsden until the tunnel was finished. Outram was asked to report, but it seemed impracticable, and no action followed.

An energetic and versatile man now appeared, John Rooth of Manchester, who was to make a great difference to the canal company's future. Accompanied by a delegation from the Ashton company, who wanted to discuss 'establishing and carrying on a thoroughfare Trade from Hull to Manchester',[13] he proposed to carry on a 'great trade' on the Ashton and Huddersfield canals if the latter would build better warehouses at each end of the tunnel, lend him £500 to start, repayable at £50 p.a. after the first year, and give him a 5 per cent drawback on tolls. They agreed, and the Ashton company also consented to lend him £1,500, to be spent on building a warehouse in Manchester, and repaid at £150 p.a., with a 10 per cent drawback on tolls because he would have to pay ground rent.

The company then decided to create 1,666 new shares,* to be offered at £30 to old subscribers and then to new. Hopefully, they sought to let a tunnel contract, but found no takers, and thought of employing John Varley, who was doing some work for them on flood repairs. They satisfied themselves of his character, but his engineering reputation was not of the best and all his sureties backed out, so after using him for some tunnelling, he was stopped from further work. Matthew Fletcher with his mining experience was then consulted. He advised that for an extra £8,000, the tunnel could be cut outwards for 1,178 yd from Redbrook pit, which was served by the big steam engine, to meet the cutting from each end. This would, he estimated, save two years of work, and the company agreed. At this time, 3,475 yd was still uncut and another 1,000 yd only partially.

In February 1801 John Rooth was asked to superintend the canal on the Ashton side, and was allowed to hire six boats from the company at 10s (50p) a week each for his carrying business. By April he had been made superintendent of the whole concern at £300 p.a. By October he was also in charge of tunnelling by direct labour at both ends, helped by John Booth. The original contractor, Lee, was now taken off and compensated. Almost simultaneously with this last appointment, Rooth asked for a road leading to the turnpike from Standedge to Oldham for his carrying business. In April 1802 the company met his road request by renting the tolls of the turnpike road from Uppermill to Standedge for three years at £160 p.a. But tolls remained negligible, £1,294 for 1802, £964 for 1803, £1,278 for 1804, and £1,322 for 1805.

The committee tried to find a contractor for the middle section of tunnel, but the only offer they had, from Jonathan Woodhouse, was withdrawn when he went to work instead on Blisworth tunnel

* 1,612 were issued, the remainder cancelled: of these 9 were forfeited. The yield was £48,190.

on the Grand Junction, and they seem then to have resigned themselves to slow penetration from each end. In the middle of 1804 the company, still having unused capital powers, decided to create 3,244 shares at £8 13s, 3,018 of which were issued and yielded £26,106.

By the end of 1804 they began to wonder how they were going to work the tunnel when they had got it, and sent a sub-committee to view 'Aircastle,* Cromford† and Harthill‡ Tunnels, and report . . . as to the time and expence it takes to pass a loaded boat through these tunnels'.[14] They came back to recommend building a towing path through Standedge, but the shareholders could not face the prospect of further delay and expense. By mid-1805, with trade improving, the company agreed to rent a wharf in Manchester from the Ashton Canal company, with whom they were working closely. Later, they bought a warehouse there.

In 1806, at a time when £308,508 had been called and £270,917 paid, another Act[15] was passed, authorizing an additional £100,000 by calls of £16 on all shares, and a special charge for passing the tunnel. The calls produced £99,808. Tunnelling went slowly on, until at last the two ends were joined on 19 June 1809.[16] Thereafter work continued on enlarging and lining, and making the approach canal and eight locks from Woolroad.

Hope was now strong, and was hardly diminished by the failure of the company's bankers in September 1810, or the bursting of Diggle reservoir on 29 November with the loss of five lives. On 1 January 1811 tolls were raised, coal, lime and stone, for instance, to 2d and merchandise to 3d per ton per mile, with an additional 1s 6d (7½p) for passing through the tunnel. Then on 26 March 1811 the great day came when the tunnel was finished and navigable, the ceremonial opening being on 4 April, when a party of about 500 entered to the strains of 'Rule, Britannia', and were watched by a crowd of 10,000. They were followed by several loaded boats, and emerged at Marsden an hour and forty minutes later.[17] On the same day John Rooth, to whom success was largely due, was directed to 'inform the public that the Huddersfield Canal is completely navigable for the conveyance of goods, wares, merchandise and all other materials'.[18]

The canal was open at last, six and a half years after its rival the Rochdale. It seems to have cost just over £400,000. About £380,000

* Harecastle, on the Trent & Mersey Canal.
† Butterley, on the Cromford Canal.
‡ Norwood, on the Chesterfield Canal.

was raised in calls, the rest borrowed or taken from revenue. Of this total, the tunnel had taken about £160,000. Standedge was by far the costliest canal tunnel built in Britain. Partly lined with brickwork, partly cut through the gritstone, 9 ft wide and 9 ft high from water level,* it was also the longest. Three passing places were provided inside it. Later, a fourth was added by the London & North Western Railway company when they became owners, and the others extended to allow more boat space and increase the number of craft that could use the tunnel at once.

Bye-laws of 1812 enacted that boats should enter the tunnel's west end between 6 and 8 a.m., and 5 and 8 p.m., and the eastern between 12 and 2, day and night. As, however, the locks were only open between 4 a.m. and 8 p.m.† in summer, and 6 a.m. to 7 p.m. in winter, boats for the night passage would have to reach the summit pound some time before they could be legged through, and wait again at the other end to leave it. In 1814 the rule was stiffened, it being said that 15 minutes after the end of the stated entrance time, stopgates would be shut and locked. Boats would then be given 3½ hours to pass. From 1817, however, boats were allowed to reach or leave the summit pound after hours if they paid 5s (25p) a time.

The canal had five important aqueducts, at Stalybridge (the one which had been rebuilt partly in iron), Royal George (Greenfield), Saddleworth, Scarbottom and Paddock. Reservoirs were built steadily, until there were ten, four of which, Haigh, Slaithwaite, Red Brook and Swellands, each held 50m. gallons. All the same, drought closed the canal several times, among them for 8 days in 1849, 35 in 1852, 23 in 1854, 66 in 1868, 41 in 1870, about 3 months in 1884, and 33 days in 1887.[19]

The tunnel open, the committee turned to encouraging trade. They observed that tolls on their own line, together with the Ashton Canal at one end and Sir John Ramsden's on the other, were higher than on the competing Rochdale, and persuaded the shareholders to cut some of the charges that had been raised so hopefully a year ago. In agreement with the Peak Forest company, who were large producers of limestone and lime, the Huddersfield cut their tolls on these and waived the tunnel charge, while the Peak Forest gave a drawback on their own charges for cargoes going beyond Standedge.[20] The Huddersfield's limekilns at Saddleworth were

* Later, it seems to have moved. In 1914 de Salis gave its then navigable dimensions as width 7 ft 8 in., height from water level 6 ft 8 in.

† Extended to 9 p.m. in 1818.

now let, with the promise of a subsidy of 10s (50p) per 100 tons on all lime over 3,000 tons a year burnt, and premiums of up to £50 each for new kilns.

The company now came up against a difficulty. The Huddersfield Canal had been built with narrow locks, to take boats off the Ashton and the Peak Forest. But these were too long for Sir John Ramsden's Canal and the Calder & Hebble, with their broad but short locks taking barges 57 ft 6 in. × 14 ft 2 in.; therefore cargoes had to be transhipped at Huddersfield. In June 1811 John Rooth was ordered to shorten two boats to pass Ramsden's Canal. This instruction originated the modified narrow boats later used on the Huddersfield, 57 ft 6 in. × 7 ft, and able to do a through journey, though carrying a smaller cargo, for in October 1812 a premium of five guineas was paid to John Harrop for building 'the first short Boat to pass upon this Canal'.[21] They were needed, for a quick result of opening the tunnel was to overcrowd the existing warehouses at Huddersfield and Cooper Bridge with goods waiting transhipment. In 1818, however, the company agreed with a group of its own shareholders to build a warehouse at Wakefield for through craft in exchange for a drawback. This was completed in 1819, and was rented by the company. Narrow boats then started to appear in Yorkshire, where existing bye-laws did not provide for them: we find a deputation from the Huddersfield visiting the Barnsley Canal in 1819 to seek 'some favourable arrangement for the Carriers in passing narrow Boats through the Locks on that Canal'.[22] The Huddersfield company were not successful, however, in representations to the Calder & Hebble on 'the propriety of building their Locks suitable for the passing of long Boats'.[23]

In mid-1812 a drawback of 1s 6d (7½p) a ton (raised in August 1813 to 2s (10p)) was given on all goods carried right through from Huddersfield to Manchester or Stockport. In mid-1813 a deputation was sent hopefully to the Rochdale Canal 'to negotiate an arrangement with them for fixing the rates of tonnage of thoroughfare goods upon the two lines upon such a scale as shall be for the mutual Interest of the two canals';[24] in other words, a price-fixing agreement. But the Rochdale thought themselves stronger, and refused. So competition drove tolls downwards, and the Huddersfield were the sufferers.

Before the tunnel opened, the canal had only been taking about £1,000 a year. Receipts then increased quickly to £6,501 for 1813, but remained at that level until they rose to £7,924 for 1819, to pause again before the £12,284 of 1822, which, though small against the £30,942 of the Rochdale for the same year, was at any rate enough to bring a first dividend within sight. It had been ten years of grind, dogged by perpetual shortage of money. In 1812 the report referred to 'the slender Capital afloat in Boats upon this Canal', and hoped road carrying would decline 'at the advanced cost of keeping Horses'.[25] In 1813 the works of the canal were said to be 'in tolerable working condition', but by 1817 it was reported that masonry work, through 'bad Construction of the Locks', would require expensive repairs for many years. John Raistrick was appointed engineer at the beginning of 1819, and as a new broom, swept clean: 'many of the locks in a most wretched condition, from the gates of some of them being much decayed, the boats having cut up the Ashler work in the lock heads chiefly from the want of Bumping-pieces to protect them, which made the leakages in many cases immense'.[26] There were also bank leakages, and some of the reservoirs were out of repair. By 1819 things were improving, and debts were being repaid, though the canal was closed for 10 days for repairs and 39 days because the reservoirs had all been emptied by the 'excessive drought'.[27] In 1820 the canal was kept open throughout the summer, in 1821 a new reservoir was begun, and in 1822, with a sharp jump in toll revenue, the committee could speak of 'this hitherto unfortunate undertaking' whose committee had, for want of funds in the past, found great difficulty in 'liquidating an accumulated Debt with Interest thereon, providing a sufficient supply of Water, and maintaining the general Repair of the Canal'.[28] The implications were that the worse was past, and better lay ahead. The total amount spent to the end of 1822 was £452,641.

Far and away the biggest traffic on the canal was classified as merchandise—in 1813 the revenue from it was £5,211 out of a total of £6,501. Next came coal, yielding £391, corn, £269, and lime and limestone, £189. But the total tonnage for the year, 40,460, was small compared to the Rochdale's 290,508. In 1816 the classified tonnages were: merchandise 16,214, coal 11,106, corn 980, and lime and limestone 3,646.

John Rooth, who had served the company so well, was given notice in 1817 so that a younger and more active agent could be recruited. He was replaced by two men, Nicholas Brown in 1814 as engineer, himself succeeded in 1819 by John Raistrick, who remained until 1843, and John Bower as agent.

Increasing traffic brought the committee to think how the 3½ hour legged passage of Standedge tunnel could be speeded up.

First, a steam tug was considered in 1816, very early in steam development. Because the company were so hard up, some committee members seem to have combined, intending themselves to run such a tug and charge each boat a fee not exceeding 3s (15p). An experiment was made in September 1817 with a smithy boat with the fire blown up with bellows to simulate a tug. It was taken through twice to put plenty of smoke in the tunnel, and then a third time with three working boats behind it. It was found that the tunnel was clear enough, and though the smithy boat's smoke was troublesome, this was 'not so much so as to endanger any person's life'.[29] The committee therefore thought that 'no great inconvenience could arise from the smoke of a small Steam Engine to haul Boats through the Tunnel'. But the idea was left for their new engineer, John Raistrick, to consider.

In 1819 he proposed a steam tug working on a chain laid at the bottom of the tunnel invert, but, though an experiment was talked of, it does not seem to have been made. Then in 1822 an engineer of the Manchester & Salford Waterworks company, Wharton, experimented with a steam boat, and as a result, tests were authorized on Raistrick's chain principle. This time they were actually made, and Raistrick himself offered to put a tug on at a charge not exceeding 3s (15p) a boat if he were allowed to operate free from dues for ten years. Presumably he did so, for a tug was working by late in 1824. By 1833, however, it had been taken off, for in June the company ruled that for the better regulation of boats in the tunnel, they would provide and pay leggers, whose charges were to be repaid by the boat captain. After 1 August no one else might leg boats through.[30] In 1834 the company thought of offering a premium of up to £105 for 'an effectual way of hauling Boats through the Tunnel', and Raistrick then suggested creating an artificial current by pumping, a device later used at Lappal tunnel near Birmingham. He was told to try.

All this time people in Ashton and Stalybridge had been putting up with towing horses dragging their ropes over local roads whose bridges were not provided with towing paths. It was not until 1834 that they complained, when the company agreed to put towing paths in tunnels under the roads at Ashton, Dukinfield, Stalybridge and Huddersfield.

After the opening of the tunnel the company's finances were very tight. They owed their bankers nearly £6,000, and proposed to seek another Act enabling them to raise £20,000 to clear their debts, and build reservoirs, warehouses and other necessary works.

But this brought such pressure upon them from millowners for still greater water restrictions that it was dropped, the company deciding instead to do without dividends until their debts had been paid off. It was not until mid-1824 that they could report the 'prosperous state of the Trade', and declare a maiden dividend of £1 per share, just over thirty years after their first Act had been passed. This and other dividends were paid on all shares *pari passu*, whatever their nominal value.* They were so exhilarated that they ordered 'that an obelisk be erected at the place where the Turnpike Road on Standedge crosses the line of the Tunnel . . . with a suitable inscription thereon', at a cost not exceeding £20, later raised to £50.†

By 1825 they were finding competition with the Rochdale line difficult, and pressed hard, but seemingly unsuccessfully, for reduced tolls on Ramsden's Canal. They also encouraged the Ashton company to make a number of useful changes: not to charge for empty boats, to allow night working, to build a warehouse at Manchester, and hoped that they would double Ashton locks, or the congestion there would encourage 'a new communication by a Rail Road or otherwise from the Huddersfield Canal to Manchester'.[31] One result was to plan a big new warehouse of five storeys at Ducie Street, their second in Manchester. This and reservoir building meant that dividends, paid for 1824 and 1825, had to be omitted until 1830. Competition with the Rochdale continued, and Sir Charles Ramsden continued to be unhelpful about toll reductions until his death. His executors in 1843 did agree to reductions, but it was then too late. In 1825 the company had large warehouses at Manchester, Woolroad and Wakefield, and smaller ones at Stalybridge and Marsden. In 1827 one was opened at Engine Bridge, Huddersfield.

Railways were first mentioned to the company in a note from their clerk about a proposed 'Rail or Tram Road from Manchester to Liverpool', about which they took no action. But at the end of 1830 there came an indication of more competitive conditions, when a carrying firm asked for toll reductions because the Trent & Mersey had given them for long hauls between Manchester and London. The Huddersfield company, itself now interested in long hauls to Goole and Hull, asked the Aire & Calder, Calder &

* Samuel Salt in 1842 gives the company 6,238 shares of an average nominal value of £57 6s 6d (£57.32½).

† If this was in fact erected, it was probably destroyed when what is now the A62 road was built in 1839 in a cutting across the earlier road.

Hebble, Ramsden's and Ashton concerns if they would consider a general reduction. This succeeded in effecting a cut in charges on timber from those ports, and in salt off the Macclesfield Canal. In 1833 all tunnel dues were abolished.

In 1835 the Aire & Calder and the Calder & Hebble called a meeting with the Huddersfield, Rochdale, Ashton and Ramsden concerns to discuss common interests, especially 'with reference to a Railroad competition', to protect their revenues, and to render their lines 'the best and cheapest means of Conveyance for general Merchandize'.[32] The result was a cut in textiles tolls to Hull, salt, and warehousing charges.

Business was now good. Toll receipts for 1838 were reported as a record, and the company raised the banks of two reservoirs to provide more water. But by June 1840 they were wondering what cuts would be needed to meet competition from the Manchester & Leeds Railway, and at the end of the year reduced many tolls by ½d a ton. Further cuts on whole-length tolls followed in the spring. In June 1841 they reported 'a most determined Railway Competition',[33] and agreed not to print and circulate accounts in future 'in accordance with the practice of various other canal companies'.

Here is the company's short dividend history:

Year	*Dividend per share*	*Year*	*Dividend per share*
1824	£1	1835	30s (£1·50)
1825	£1	1836	35s (£1·75)
1826	none	1837	£2
1827	none	1838	£2
1828	none	1839	£2
1829	none	1840	£2
1830	10s (50p)	1841	£2
1831	£1	1842	none
1832	£1	1843	none
1833	30s (£1·50)	1844	none
1834	30s (£1·50)		

After 1841 toll cuts continued, economies of £2,000 p.a. were made, an 'unparalleled depression of trade throughout the whole District traversed by the Canal' was added in 1842, and no further dividends could be paid. More traffic was obtained but 'it has not yet been attended with an adequate improvement of Income'.[34]

Then, in September 1843, Captain Laws of the Manchester & Leeds Railway made it known that he was willing to meet the company's agent 'for the purpose of endeavouring to effect some arrangement between the two Companies'.[35] The company agreed to meet him, and the railway company then said that they would suggest figures for a fixed-period lease.

In 1844, however, the Huddersfield & Manchester Railway was projected to build a line from Stalybridge on the Sheffield, Ashton-under-Lyne and Manchester Railway to the Manchester & Leeds at Cooper Bridge. The Manchester & Leeds had also suggested a branch of its own from Cooper Bridge to Huddersfield, for the *Railway Times* of 9 March 1844 describes a public meeting in opposition, which calls that company the 'most brutal line in the whole kingdom', but this was withdrawn.[36]

Agreement was reached with the Huddersfield & Manchester in May 1844 to buy the canal. Its name was changed to the Huddersfield & Manchester Railway & Canal Co, and it was authorized in 1845[37] with a capital of £630,000, and power to increase it to buy the canal company at the rate of one £30 railway share for each canal share. Those who were not willing to make the exchange were to get £25 cash. Under this Act 5,552 railway shares were issued for canal shares, and 687 canal shares were paid off in cash. The new company therefore paid £183,730 for the canal. After the share exchange had been completed, the canal company was dissolved.

In early 1845 the Sheffield, Ashton-under-Lyne & Manchester Railway agreed to lease the Huddersfield & Manchester, now beginning to build its railway along the canal line. The lease was approved by the former's shareholders, but rejected by those of the Huddersfield & Manchester who preferred independence to becoming part of what was to become the Manchester, Sheffield & Lincolnshire Railway and later the Great Central. However, in 1847 another Act enabled them to lease themselves in perpetuity to the London & North Western Railway, which they did, shareholders receiving seven-tenths of the dividends paid on shares in the parent company.

A new schedule of canal tolls was laid down, most goods being charged at 1d or less, and the Commissioners of Railways were empowered to require the London & North Western Railway 'to proceed to the Correction or Prevention of any Inconveniences or Evils' resulting from the transfer to them of the Huddersfield or Ramsden's Canal. The new railway was opened in 1849 through a railway tunnel at Standedge, and connected at Stalybridge with the Manchester, Sheffield & Lincolnshire.[38] (*To continue the history of the Huddersfield Canal, turn to p.* 445.)

CHAPTER XIII

The Carlisle Canal

THE River Eden runs from the Solway Firth past Carlisle and then south-eastwards through Cumberland. Before 1720, there had been a trade by sea from collieries in the Maryport area to places up the river, but when the coastwise duties were imposed, it was found cheaper to carry coal to Carlisle and its neighbourhood by land, and the river trade ended.[1] An Act was then sought by three Carlisle men, Thomas Pattinson, John Hicks and Henry Orme, to waive the coastwise duties between Ellen Foot (Maryport) and Bank End in the river itself, build wharves, cranes and warehouses, and dredge as necessary, for the Eden was then

> 'so very shallow in several Places thereof, that Boats, Lighters, Barges, or other Vessels, cannot pass up to . . . Bank End . . . except it be in times of Flood or High Waters, which occasions great want of Fuel, and other Necessaries, for the Supply of the Inhabitants'.

The preamble went on to argue of the river that:

> 'several Vast Tracts of Ground lying waste and uncultivated for want of water-carriage of Coals, Lime and other Manure for that Purpose, renders it impossible for the said Inhabitants to burn Lime sufficient to lay on and improve the same, the Improvement whereof would employ, relieve and be very advantageous to the Poor, who at present, are a very great Burthen to the Parishes near and adjoining to the said River'.

The Act[2] passed in 1721 empowered them for 31 years to levy tolls of 8d a ton on coal, lime and limestone, or 1s 4d (6½p) on cinders, on goods landed between Ellen Foot and Bank End, and remitted the coastwise duties, but gave no authority to make cuts or locks. Presumably the aftermath of the South Sea Bubble prevented action.

Some years earlier, on the other side of England, there had been a parliamentary petition in support of extending the navigation of the Tyne upwards from Newburn to Hexham, but this also came to nothing. Except for a Tyne scheme in 1778, there were no further proposals for waterway improvement in Cumberland or Northumberland until 1794, when they began to pour out. Most were for extending or by-passing the Tyne, but were presented as parts of possible canal lines from Newcastle to Carlisle or Maryport.[3]

Carlisle men never supported these sea to sea projects as enthusiastically as did those of Newcastle. Their interests were more local, in improving facilities for coastal craft from Liverpool, Ireland and Scottish ports trading to their city. In 1795 three craft averaging 35 tons burthen were running between Liverpool and Sandsfield* on the south bank of the Eden below Carlisle, carrying about 1,000 tons a year. At Sandsfield the cargoes were transferred to road waggons for the 3½ miles haul to the city.[4] In 1807 William Chapman noted that 'Liverpool flats or canal boats frequently come from the Mersey, mostly laden with fir timber'. While waiting for a suitable tide to Sandsfield, craft from Liverpool, Glasgow, Whitehaven and other ports were accustomed to lie off the land between Knock Cross (just above Bowness) and Glasson Lane Foot a little higher up the Firth. By 1818 it was said that 'there are at this present time . . . six trading vessels regularly passing between Liverpool and Bowness or Sandsfield, beside occasional ones'.[5]

They started to take action in 1807 when, after a public meeting on 21 May, a committee was formed in Carlisle to promote a canal thence to the sea mainly to provide the city with a better coal supply. They asked Chapman to report.[6] He maintained his 1795[7] preference for Maryport with its existing harbour and coal supplies as a terminus, but thought that if the likely cost of some £90,000 to £100,000 was too great, a terminus east of Bowness on the Solway Firth would be practicable. A canal to take 45 ton craft would cost about £40,000, but if it could be enlarged to take 90 ton vessels at about £55,000 to £60,000, these could trade to the Forth & Clyde Canal or across the Irish Sea. Such a canal would be also a first step towards a sea-to-sea waterway.

His report was considered at a committee meeting in Carlisle on 29 June 1807, which ordered subscribers to be asked whether they preferred a Maryport termination, or one on the Solway Firth on

* New Sandsfield, where the road ended and goods were landed, was then sometimes called Port Carlisle. It is so named on the Carlisle Canal deposited plan of 1818. This must not be confused with the later Port Carlisle at Fisher's Cross near Bowness.

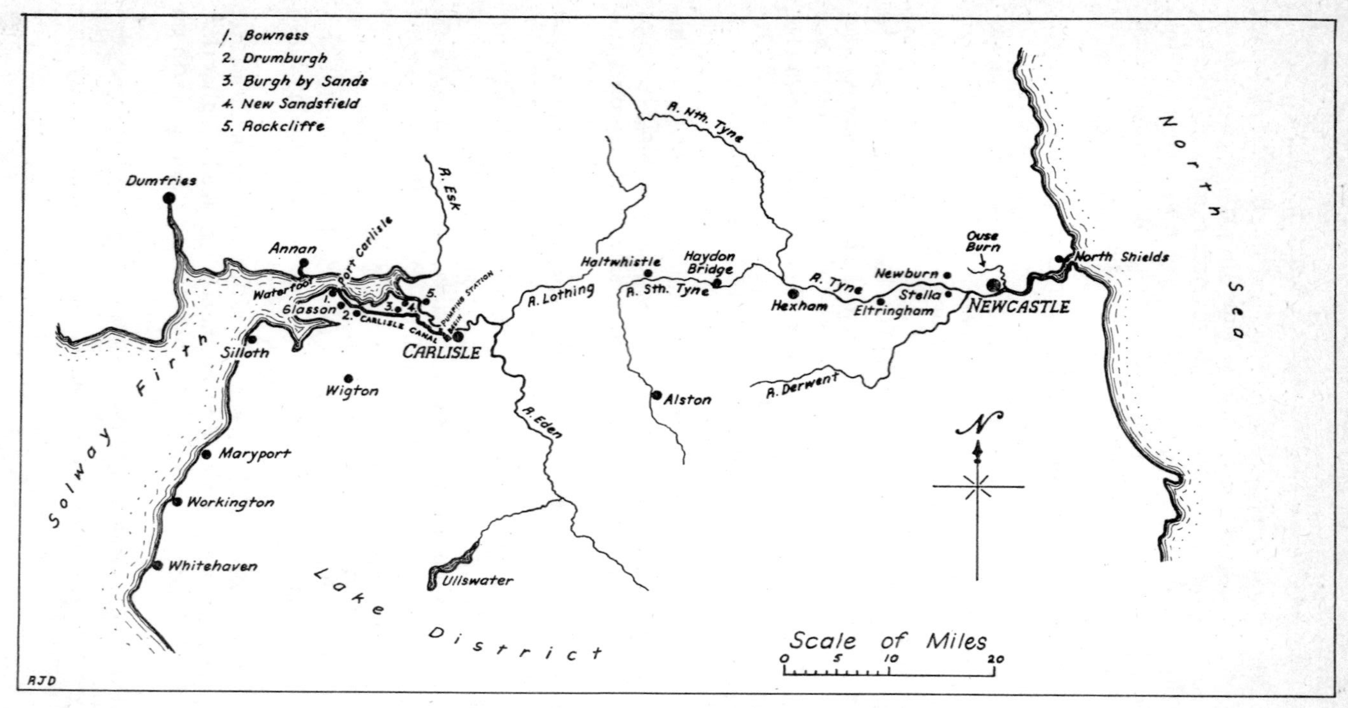

24. The Solway–Tyne area and the Carlisle Canal

a large or a small scale. On 22 August Chapman added an appendix[8] to his report, proposing both a 90–100 ton canal for the Irish, Scottish and Liverpool trade, and a 50 ton line to the collieries near Maryport, and said: 'I see by the newspapers, that the Sea Vessel Canal from Carlisle appears to be as warmly espoused as that to Maryport for Canal Boats.'

Thomas Telford was then called in, and on 6 February 1808 reported on what he called the Cumberland Canal, saying that his plan was based on uniting Carlisle with other parts of the country that would benefit by water transport, while at the same time making what was done consistent with a future sea-to-sea canal. From Carlisle's point of view, he said, seagoing vessels should be able to reach a basin near the city where they could load and unload under the eyes of merchants and manufacturers, and where farmers could deliver produce and receive lime.

He proposed a canal not smaller than the Forth & Clyde, with locks 20 ft wide and 8 ft over the sills to take coasting vessels, from the Solway Firth about a mile east of Bowness* to Carlisle at a cost of £109,393, with a navigable feeder to Wigton on a smaller scale, probably taking 7 ft wide narrow boats, for £38,139 more. The main estimate could be reduced, he said, by £20,000 if the canal were made the same size as the Bridgewater, or by £50,000 if for narrow boats. But both would involve transhipment. Such a canal, he added, could later be extended to Eskdale, and up the valley of the Eden as part of a line to Newcastle.

Chapman, commenting on Telford's proposals, sensibly thought a steam engine a cheaper way of supplying the canal with water than the Wigton branch unless the latter could stand on its own feet commercially, and that locks could be made rather smaller than on the Forth & Clyde—perhaps 65 ft × 16 ft × 6 ft to take 100 ton craft, which would be easier to tow. Such a canal, he said, would be big enough for Mersey flats, which would be likely to work the Liverpool part of the expected trade.[9]

The idea was then laid aside, to be taken up again in the autumn of 1817. A meeting at Carlisle on 7 October asked Chapman to survey a canal from Carlisle to the Solway Firth for craft of not less than 70 tons, and told him that he should 'strictly adhere to the great ultimate object of connecting the east and west seas'. He recommended[10] a canal from Fisher's Cross† to Carlisle, 50 ft wide and 8 ft deep, with locks 74 ft × 17 ft, at an estimated cost of

* Roughly where Port Carlisle later was.
† Later to be called Port Carlisle.

£73,392. The committee themselves forecast a revenue of £7,365 p.a. A smaller-scale canal, Chapman said, could be continued to Newcastle, and also up the vale of Eden to the slate quarries by Ullswater. His report was accepted, money was raised from such local notabilities as Lord Lonsdale (£5,000), Lord Lowther (£1,000) and Sir James Graham (£2,000), from Carlisle corporation (£1,000) and from leading citizens, and an Act[11] obtained in 1819 for the Carlisle Canal which authorized a capital of £80,000 in £50 shares, and £40,000 more if necessary.

At their first meeting after the Act, the committee led by Dr John Heysham their chairman, very sensibly decided that before they started work, they had better look at some other canals; deputations were therefore sent to see the Forth & Clyde and the Lancaster. William Chapman was taken on as consulting engineer, and soon afterwards Henry Buck, the brother of Richard who had helped Chapman with the preliminary surveys, arrived to learn what Richard could teach him and to take over as resident. By early 1820 contracts were let for the whole canal, and then in March the committee took on Thomas Ferrier from the Forth & Clyde Canal to be overseer of works, seemingly because they were not getting on well with Chapman and Buck. This move led to Henry Buck's resignation as resident in July, after which Richard seems to have stayed on, working to Ferrier who now in fact acted as site engineer. This annoyed Chapman, who in November 1822, when most of the construction had been done, reported severely on some of Ferrier's work, and recommended that Richard Buck should finish it. The committee disagreed:

> 'Mr Chapman's repeating such recommendation in opposition to the opinion of the Committee at various times expressed to him and notwithstanding the many difficulties which now appear in the Work and which have in a great degree arisen from the Want of an active and vigilant Superintendance does not by any means accord with that attention to the Interest of the company which they have a right to expect from their chief Engineer.'[12]

They therefore dismissed Chapman,* and Buck was asked to leave two months later, just before the canal opened on 12 March 1823. He did not, however, do so until about May. Ferrier remained until the end of 1826.

The canal was 11¼ miles long, 54 ft wide at surface and 8 ft 6 in. deep, with locks 18 ft 3 in. wide and 78 ft long. From a wooden jetty built on the seaward side of the canal entrance at Fisher's

* Chapman later published a pamphlet giving his point of view.[13]

Cross, now renamed Port Carlisle, it rose through the entrance lock and another, and ran level for nearly 6 miles; it then rose by 6 locks in 1¼ miles, and again ran level to Carlisle basin, some 450 ft × 120 ft. Soon after opening, a three-storey brick warehouse was built at the far end of the basin (it is dated 1823), and on the east side a long low building with access from a roadway on a lower level, for coal and lime storage. In 1838 a timber pond was excavated on the inside of the curve the canal made as it approached the basin. The entrance lock was built level with high water at lowest neap tides, and the long level above the second lock was 6 in. above the highest known tide. The canal had no overbridges, only two-leaved drawbridges built to the Forth & Clyde pattern, so that it could be used by coasting craft. Of the £80,000 authorized in £50 shares, £70,600 had been subscribed, mostly by local interests, many in the textile industry. Some £10,000 had also been borrowed by the time the canal opened. The cost to this time was therefore some £81,000.

The committee worked to get a commercial trade started. They rented a stone quarry; a timber yard was built at Carlisle and timber rafts used the canal. Then the Treasury agreed that the coasting duty would not be charged on coal, stone and slate carried between Whitehaven and Carlisle, but the committee had difficulty in finding anyone willing to begin a coal trade, and in June themselves sent the *Mary* to Harrington to get a cargo. They sold this, but in October minuted that they would not take part in the coal trade, but leave it 'entirely to the Competition of Individuals'.[14] Trackers* established themselves to provide towing services, and by the end of the year at least one vessel in the Irish trade had come to Carlisle, and the company had taken £928 in tolls. In 1824 they got another trade started, this time bricks, by importing two boat loads for public sale. Business was increasing, and in spite of having built a reservoir they foresaw future water shortage. They therefore cut a feeder from the river, and built a waterwheel to lift it to the canal. Some ships, however, especially in the timber trade, avoided paying canal tolls by coming up river on favourable tides to Rockcliffe on the north bank above Sandsfield, where road waggons met them.

The committee had from the beginning looked forward to a steamer service from Liverpool to Port Carlisle and a packet boat on the canal running in connection with it. Therefore when in 1825

* Boatmen usually provided their own towing horses, but when many sailing or seagoing craft used a canal, then towing services were provided either by independent groups of men, called trackers, or by the canal company itself.

the Carlisle & Liverpool Steam Navigation company asked for an exclusive berth at Port Carlisle for their vessel, the canal company agreed to build it, cost to be repaid over ten years with interest. They also laid out their jetty afresh, and looked about for a packet boat. They found the *Bailie Nicol Jarvie*, bought it second-hand from the Edinburgh & Glasgow Union company, and let it to Alexander Cockburn, a local innkeeper, for £30 p.a., and afterwards to various lessees. It began running on 1 July 1826, when a 'good number of Passengers had a delightful trip listening to the band provided, enjoying the hospitality of the said Mr. Cockburn and inhaling the sea-air'. About this time the steamer started running, so that it became possible to leave Carlisle in the morning and be in Liverpool that evening, though the packet-boat services seem at first to have been worked only in the summer. Goods were also carried by the steamer, the Steam Navigation company possessing lighters in which they were taken along the canal. Perhaps now, perhaps a little later, the Solway Hotel (now Solway House) was built at Port Carlisle for the entertainment of transit passengers.

In August 1824 meetings were held at Newcastle, and a committee formed, to promote either a canal or a railway thence to Carlisle. William Chapman, impressed by the Stockton & Darlington Railway and the threat it offered to Newcastle's coal trade, had written a pamphlet[15] to explain that his 1796 line of canal could also be used for a railway. He was therefore asked to estimate, and on 27 October 1824 reported the respective costs would be £888,000 and £252,488. Rather naturally, in March 1825 the decision was taken for a railway, and a company was formed, though the Newcastle & Carlisle Railway's Act was not passed until 1829.[16] The company had been in touch with the canal committee during 1825 about ending their line at the basin, and had been welcomed, for such a connection could only increase the canal's trade. Indeed, the shareholders' meeting of July 1828 minuted that it was 'of the utmost importance to the interest of this Company that the proposed Railway between Carlisle and Newcastle should be carried into effect and that every possible exertion should be used in Carlisle and the neighbourhood to promote that object'.[17] John and George Dixon and William Forster, canal committeemen and substantial shareholders, supported the railway also, and were directors in 1835.

The railway was opened in sections, those most important to the canal company being Rome Street (Carlisle) to Blenkinsopp colliery, Greenhead, 19 July 1836, Canal Basin to Rome Street, 9

March 1837, throughout to Redheugh (Gateshead) on 18 June 1838 and into Newcastle in 1839. This line very greatly helped the fortunes of the canal company, for on it moved goods not only from the north-east coast but also from the Continent for the north-west and Ireland, and also passengers, notably German emigrants on their way to Liverpool and so the United States.

Here are averaged figures for canal tolls (excluding packet-boat takings) for the first twelve full years;* these are followed by single-year figures for 1836 to 1840 to show the difference made by the railway to commercial and packet-boat receipts and to dividends:

Years	*Average tolls* £	*Average dividend per cent*
1824–26	1,659	–
1827–29	2,305	–
1830–32	2,488	–
1833–35	2,905	1

Years	*Tolls* £	*Packet-boat Receipts* £	*Dividend per cent*
1836	3,287	578	1
1837	4,103	625	1
1838	4,915	570	3
1839	6,110	645	4
1840	6,605	829	4

The company's first dividend, of 1 per cent, was paid in 1833. In the ten years since opening, they had borrowed and spent some £23,000 upon improvements in addition to their former £10,000 debt, and in 1834 the shareholders authorized another £5,000.

From 1832, with the railway being built, the canal company set their minds to what would happen when the line opened. Early in that year a committee of shipowners who had earlier arranged a buoying of the Solway Firth channel and the collection of funds from ships to repay expenses asked the company to take on the job and collect contributions. They agreed. A year later, in August 1833, a deputation arrived from a second steam navigation company, the Carlisle & Annan (later, the Carlisle, Annan & Liverpool) which proposed to put a vessel on the Liverpool run, starting from Port Carlisle and calling each way at Annan. Again the canal company consented to build a berth for it.

* The figures are for approximate calendar years, depending upon the dates of committee meetings.

These augmented steamer services and the prospective arrival of passengers by rail at Carlisle from the east suggested a much faster packet boat on the canal. The committee therefore asked William Houston of the Glasgow, Paisley & Ardrossan Canal[18] to oversee the building for them of a lightly built swift boat of his new design. This, the *Arrow*, started running early in the summer of 1834, being now worked directly by the company, with contributions towards tracking costs made by the steamer companies. Starting with the winter of 1836–7, the packet ran all the year round, instead of only in the summer, helped by the new icebreaker the company had had built for them. The *Bailie Nicol Jervis* retired and was sold for £7 12s (£7·60). That year they also started to run an omnibus to take passengers from the basin into Carlisle itself.

With the prospect of more traffic, water supply was likely to be a problem. Instead of more reservoir capacity, in 1834 the company decided to install a new waterwheel and pumps. William Fairbairn built them for £1,391, and they were working in the following year. Pump efficiency was lower than Fairbairn had forecast, however, and in 1838 Harvey's of Hayle supplied a 60 in. steam engine and pumps for £3,700. This was only used when the river was too low to use the waterwheel, or when the reservoir needed filling.

Early in February 1835 the Admiralty, already surveying at Liverpool, were asked also to survey the Solway Firth and Port Carlisle, and agreed. The railway was now almost finished, and the pace increased. Shipowners were asked to increase contributions towards keeping up the Firth buoys, the ship berths at Port Carlisle were dredged, and in September the first plans were made for a Bill to authorize interconnected inner and outer docks at Port Carlisle. By November Jesse Hartley of Liverpool had produced plans which, with a draft Bill, were approved early in 1836. The enabling Act,[19] which raised the company's borrowing powers by £40,000, and empowered them to light and buoy the estuary, was passed that year, and in June Hartley proposed a start 'by inclosing the proposed Extent of Dock Ground and completing in the first instance only the outer Dock with single Gates'.[20] However, Lord Lonsdale as Lord of the Manor had rights over the foreshore, and, though £25,000 had been borrowed, work did not start until August 1838. Meanwhile the Admiralty were surveying the Firth, and their advice was asked about lighting and improved buoying. Eighteen buoys were ordered in May 1837 and installed a year later; in 1840 a lightship was built (it started work the next year) and a lighthouse built and manned at Lees Scar off Silloth, costs being recovered by tolls of 1½d a ton on shipping. Competition was coming, however, for the Maryport & Carlisle Railway, offering a better port than Port Carlisle, was authorized in 1837.

The railway connection to Carlisle canal basin led to an increase of traffic, especially in coal from Lord Carlisle's mines and those of the Blenkinsopp Coal company at Greenhead. At first the canal company considered boats or rafts to carry railway waggons, of the kind used on the Forth & Clyde or the Don, but decided against them and in favour of coal barges hauled in the Firth by a steam tug. In 1838 and 1839 canal and railway rates were both reduced to encourage the coal and sea-to-sea traffic by rail and canal. Meanwhile gratuities were paid to lock and bridge keepers for their extra work in passing traffic. In July 1838 a second swift packet boat, the *Swallow*, arrived from Paisley to improve the passenger service.

In October 1838 the two steamer companies discussed with the canal committee a proposed new company to run packet boats on the canal and also passenger steamers and tugs on the Firth. This idea was not followed up, but out of it came a canal company decision themselves to start a passenger and goods steamer service across the Firth from Port Carlisle to Annan. In December the *Clarence* was bought second-hand in Glasgow for £1,200, and after a jetty had been built on the Annan side, she started running in 1839, carrying passengers and towing lighters.

From early 1841 severe competition developed in the export of coal, presumably from Maryport and farther south, and in April the committee were told that its diminution or entire failure could only be prevented by cuts both in canal and Newcastle & Carlisle Railway tolls. These were made, and had a serious effect on receipts, which fell by nearly a quarter until they recovered again in 1845.

By the end of 1844, the worse threat was looming of the completion before long of the Lancaster & Carlisle Railway, then building a line which, when finished, would link Carlisle by rail with Liverpool and Manchester and threaten the canal and steamer route. A few months later, a further threat came from the 'projected railway from Glasgow by Dumfries and Annan to Carlisle',[21] the Glasgow, Dumfries & Carlisle, but the company could only oppose and unhopefully seek compensation. At this time the canal shareholders thought it best to get rid of their debt of £77,437 10s by apportioning it among the shareholders at the rate of £52 10s per share. Shareholders could either pay off this sum in cash, or have interest at 4½ per cent deducted from their future dividends. Only £1,522 10s was liquidated by 8 shareholders. When dividends ceased those who had paid their share of the debt were given the appropriate interest.

The events of these years must be looked at against the history of the Maryport & Carlisle Railway. In 1836, while the Newcastle & Carlisle was still building, a company was formed to construct it, at first to Whitehaven, but later on financial grounds only to Maryport. Such a line would link Newcastle with the west coast and so with the Irish trade, and also add to the market for coal from the northern part of the West Cumberland field round Maryport. George Stephenson was taken on as engineer, and the line authorized in 1837. But money was difficult to raise, John Blackmore the Newcastle & Carlisle's engineer replaced Stephenson late in 1839, and it was not until 10 February 1845 that the line was opened throughout. The separate Whitehaven Junction Railway was opened from Maryport to Whitehaven for goods on 23 February and passengers on 19 March 1847.[22] The opening of the Maryport & Carlisle Railway does not seem to have affected the canal tolls until worse threats came from lines opening to Carlisle from both north and south, and had a cumulative effect from 1847 onwards.

In July 1845 the Newcastle & Carlisle Railway and the canal increased their coal tolls, and at once released threats from shippers to use the Maryport & Carlisle Railway instead. A study of Maryport & Carlisle charges then brought a reduction in timber tolls and harbour dues, and also in spring 1846, a successful request to Trinity House to license two pilots to be stationed at Maryport but paid by the canal company, to bring vessels to Port Carlisle. These were at work by the autumn. The full length of the Maryport & Carlisle was finished on 10 February 1845, the Lancaster & Carlisle on 17 December 1847, and the Caledonian on 15 February 1848. But by the autumn of 1846 the steamer companies were asking for reductions in dues, and the canal company had sought a preliminary report on converting its own line to a railway. This, received in February 1847, was favourable, and in April the committee resolved: 'That Mr Errington be requested to make such a Survey and Report for the purpose . . . as will enable the Company to obtain a guaranteed Tender for the Work.'[23] A month later the directors of the Newcastle & Carlisle Railway arrived to view the canal, and were given a dinner at Port Carlisle. Presumably they were wondering whether to participate in the conversion. In July a shareholders' meeting told the committee to 'open a negotiation for connecting this undertaking with the Newcastle and Carlisle

Railway Company or any other of the Railway Companies connected with Carlisle'.[24]

For a time, however, railway conversion plans were laid aside, and the coal concerns, the steamboat companies and the canal set themselves to reduce expenses and cut tolls. Leasing the canal was thought of, but the only offer they had, from local people, was not enough. In September 1848 the company heard that George Hudson, energetic chairman of the York, Newcastle & Berwick railway, was negotiating to lease both the Newcastle & Carlisle and Maryport & Carlisle railways, and minuted that should he do so, they should try to prevent him undercutting them on the Maryport line on articles usually carried by the canal. The lease took effect on both lines on 1 October, but the shareholders of the York & North Midland refused to accept what their chairman had done. In 1849 Hudson fled abroad, and on 1 January 1850 the two lines again became independent.[25]

So far passenger traffic had kept up reasonably well, though the *Clarence* had been sold early in 1847 and the Annan service ended. Some was lost to the Lancaster & Carlisle and Glasgow & South Western railways when they opened, and more when in April 1850 cheap steamer fares were introduced on the route from Carlisle to Whitehaven by rail and thence to Liverpool by a shorter sea passage than from Port Carlisle.

The autumn of 1850 saw the last interest payment made on loans, and still more toll cuts on most commodities, coupled with efforts to persuade the Newcastle & Carlisle Railway not to favour the Maryport & Carlisle, and Newcastle merchants to use the canal to import slate. It was hopeless, and in March 1852 the committee minuted: 'it appears highly desirable and indeed the only means of retrieving the affairs of the Canal Company that the Canal should be converted into a railway'.[26] Their engineer was told to plan conversion and extensions at Port Carlisle, and the Newcastle & Carlisle were asked to allow theirs to help him. Money was raised mainly from share and loan holders, and a Bill prepared, the committee resolving that the conversion line should be built,

> 'with a view to the ultimate extension of the Railway to Silloth Bay as the natural and in every respect the most desirable terminus of the line and the most eligible Port for the City of Carlisle as appears from the investigation made by this Company in the year 1847'.[27]

If trade were not to be permanently lost, conversion had to be done as quickly as possible. Even before the royal assent to the Act on 3 August 1853, work had begun in June, and the canal had been closed on 1 August, an omnibus meanwhile taking passengers from Carlisle to the steamers. With the Act, the canal company ceased to exist, being replaced by the Port Carlisle Dock & Railway company. Here are tolls, passenger receipts, harbour dues and dividends to the closing date:

Years	*Tolls*	*Passenger Receipts**	*Harbour dues*	*Dividends per cent*	
	£	£	£	£	
1841	4,785	800	434	3	
1842	4,774	794	364	–	
1843	4,603	746	369	1	
1844	4,872	803	399	1½	(1·50)
1845	6,349	864	344	2½	(2·50)
1846	6,300	887	381	3-13-6‡	(3·67½)
1847	5,418	580	321	4‡	
1848	4,090	590	319	3-2-3‡	(3·11)
1849	3,538	526	259		
1850	3,177	367	215		
1851	2,209	329	135		
1852	1,781	330	168		
1853†	1,316	208	96		

* From the packet boats, but not the *Clarence*.
† To closure on 1 August.
‡ After apportionment of debt.

The capital position of the canal company as stated in the Act was as follows: £76,600 had been raised in shares, of which 1,412 were in issue, 63 having been forfeited. The debt was £44,370, plus £27,645 under the dock Act.

The line was opened to Port Carlisle for goods traffic on 22 May 1854 and for passengers on 22 June. However, in 1853 a prospectus had been issued for the Carlisle & Silloth Bay Railway & Dock company, with a capital of £165,000, to build a dock at Silloth to rival Maryport, and a line to join the Port Carlisle Railway at Drumburgh. The Act was passed on 16 July 1855, and the line opened in 1856. Both lines were leased to the North British Railway in 1862 and merged with it in 1880. The Port Carlisle branch, as it now became, was in its time well known for its four-wheeled horse-drawn passenger 'dandy', used from 1856 to 1914. The branch was closed on 1 June 1932. Those with a seeing eye can still find many traces of the canal, from the warehouse and coal stores at Carlisle basin to the old entrance at Port Carlisle.

The Nent Force Level

The Nent Force Level[28] was a mine drainage tunnel nearly 5 miles long beneath the valley of the River Nent from Alston to Nenthead, in east Cumberland. Alston Moor and land around it was owned by the Commissioners of Greenwich Hospital and the leasing of mining rights was administered by two Receivers who, in 1775, were John Smeaton and Richard Walton, the year in which they prepared a report and a plan of the proposed level. It was intended to intersect a number of lead ore veins which, without drainage, would be impossible to work economically. Smeaton appears to have acted only in a consultative capacity, but Walton was more intimately connected with the work which started on 10 June 1776 at Alston.

It seems that it was probably intended to use the level as an underground canal from the start, in order to boat out spoil, but as the dimensions were minimal—only 3 ft 6 in. wide—it was found easier to cart it out. In 1777 the Commissioners were advised to increase the width by John Gilbert, the Duke of Bridgewater's agent, 'that it may serve as a navigable Canal; in order that it may be seen whether the expense of making the said Canal will not, as he represents, be greatly different from that of making the said level of the size already begun'. Gilbert had local connections, having contributed to a mining venture in Middle Fell between the South Tyne and the Nent in 1771. Walton and Smeaton, too, were involved, proposing to drive two underground canals, one from each river, to drain and work the mines. A shaft was sunk for the South Tyne venture but was abandoned on striking hard basalt rock.

The Nent Force Level was accordingly widened to about 9 ft square, but progress was very slow and it took until 1810 to reach Lovelady Shield Shaft, some 2 miles 921 yd from the Alston portal. Here the policy of driving a horizontal level was abandoned in favour of an inclined one, commencing 210 ft up the shaft from the original level. Consequently navigation stopped at this point. The 'level' was finally completed to beneath Nenthead in 1842, and with a later extension totalled 4·94 miles in length.

In the first half of the nineteenth century an underground voyage on the level was one of the tourist attractions of Alston. The water was some 4 ft deep and was navigated by 30 ft boats 'propelled by means of sticks projecting from the sides of the level'. Lighting was by candle and apparently musical accompaniment was a feature of the journey on occasions. Expeditions were organized by a local innkeeper, and near the end of the navigable portion dances were held on a smooth rock platform just above water level. Boats were kept at the Alston end by the mining company until at least 1900, perhaps longer, but mining virtually ceased in Alston Moor in 1951 and now all traces of the entrance to the level have disappeared.

PART THREE—1845–1969

CHAPTER XIV

The Bridgewater Undertaking

WE can distinguish three periods in the history of the Bridgewater trustees before their undertaking was bought in 1872 by the Bridgewater Navigation Company: that from the situation described at the end of Chapter V to 1851, of heavy fighting against both railway and other canal competition; the four following years to 1855, when they worked for an alliance with the Great Western Railway; and the period from 1855 to 1872 when, Loch having died, Fereday Smith the general manager and the Hon Algernon Egerton as superintendent reformed and developed the business, notably by improving facilities at Runcorn, Liverpool and Manchester. As a background to all this, from 1844 onwards Loch had a possible sale of the trustees' waterways in mind, whether for conversion to a railway or a ship canal, or as part of a railway-controlled whole, his main motive being to safeguard Lord Francis Egerton's (later Lord Ellesmere's) future income. The intricate story of these phases has been told by Mr F. C. Mather in his new book, *After the Canal Duke*. The following account, therefore, should be regarded only as a brief introduction to Mr Mather's work.

The rates agreement of 1845 between the trustees, controlling the Bridgewater and Mersey & Irwell water lines, and the Liverpool & Manchester Railway, lasted only a few years, and produced the following figures[1] of tonnages carried:

Date	*Waterways* *tons*	*Railway* *tons*
1845	455,478	176,512
1846	468,845	190,745
1847	388,241	164,081
1848	407,832	151,184

It was ended by the opening of the Liverpool & Bury Railway. This line, built to attack what Bolton and Wigan industrialists considered to be the rail monopoly of the Liverpool & Manchester, was authorized in 1845 to run from the Manchester & Bolton Railway near Bolton (whence there were to be connections through to the Manchester & Leeds Railway at Rochdale via that company's Heywood branch) to Liverpool.[2] It was opened at the end of 1848, the company having meanwhile amalgamated with the Manchester & Leeds in 1846, which in turn became the Lancashire & Yorkshire Railway in 1847.

The agreement ended, price-cutting and confusion followed. In 1885, giving evidence on the Ship Canal Bill, a Bridgewater spokesman said that for two months in 1849 the trustees were carrying cotton to Manchester for 2s 6d (12½p) a ton, and manufactures back to Liverpool for 1s 8d (8½p). Passengers were taken between the two towns for 3d. In the heat of battle the trustees alleged that the railways wished to ruin the private canal carriers, and in June 1849 agreed to make the carriers their agents, themselves fixing rates and in return bearing any losses. They thus forestalled railway attempts to offer similar arrangements. This system, which led the trustees into carrying commitments over waterways far from home, outlived the bitter competition that had created it. A year earlier they had also acquired the Anderton company carrying mainly on the Weaver and Trent & Mersey via Anderton, and in 1849 leased the tolls on the Trent & Mersey's through traffic for seven years, took over the North Staffordshire Railway's* canal carrying stock at Wolverhampton, and came to an agreement with Shipton & Co, canal carriers at Wolverhampton, who were linked with the North Staffordshire Railway. They thus obtained a strong position as against the railways in the midlands trade.[3]

In January 1850 the trustees found it necessary to make a disadvantageous agreement for sharing the trade between Liverpool and Manchester on a fifty-fifty instead of the former basis of approximately two thirds-one third. By it they were 'unwillingly obliged to undertake on their own behalf, and on that of the Carriers on their Navigations, to withdraw from the Public and their Customers many accommodations and Services they had hitherto been in the habit of granting'.[4] The railways would not, however, negotiate an agreement for traffic going to, or coming from, places beyond Manchester. Much of this trade came and went by road, though the railways had lines, the London & North Western to Stockport, the Lancashire & Yorkshire to other towns.

* The North Staffordshire Railway had bought the Trent & Mersey Canal in 1846.

THE TRUSTEES

OF THE LATE

DUKE OF BRIDGEWATER

DAILY

CARRIERS,

BETWEEN

BIRMINGHAM	Great Charles Street.
DUDLEY	Tower Street.
TIPTON	Canal Wharf.
WOLVERHAMPTON	Albion Wharf.
KIDDERMINSTER	Park Wharf.
STOURPORT	Danks, Venn & Sanders.
WORCESTER	Worcester Canal Carrying Company.
GLOUCESTER	Danks, Venn & Sanders.
BRISTOL	Danks, Venn & Sanders.

AND

The Staffordshire Potteries.	STOKE	The Anderton Company's Wharfs.
	ETRURIA	
	PORT VALE	
	LONGPORT.	
	TUNSTALL.	
	NORTHWICH	
	CONGLETON	
	MACCLESFIELD	
	BURTON-ON-TRENT	
	DERBY	
	LIVERPOOL	Old Quay Dock.
	MANCHESTER	Ashton Canal Wharf

25. A trustees' directory advertisement of 1850 for their carrying business

Of the canals, the Rochdale was independent, though friendly to the Lancashire & Yorkshire, but the Manchester, Bolton & Bury was controlled by the L & Y, the Huddersfield by the L.N.W.R., and the Ashton and the Peak Forest by the Manchester, Sheffield & Lincolnshire. The railways for the time being carried at the same rates to and from places such as Oldham as to Manchester itself, while trying to exclude the Bridgewater from the Manchester, Bolton & Bury by charging maximum tolls, and from the Huddersfield Canal by restricting private carriers. 'But the Turnpike Roads and the Spinners' Carts still remain', as the trustees wrote.[5] They were determined to keep the trade, and therefore allowed traders the cost of carriage to places beyond Manchester. The agreement did not last long. There was another outbreak of price cutting in 1851, followed by another uneasy peace.

Queen Victoria rode on the canal to Worsley on 10 October 1851—unfortunately a wet day. 'The Queen reached the Patricroft

Station on the Liverpool and Manchester Railway at half-past four o'clock in the afternoon; and there embarked in the state barge, which conveyed the Queen, her consort, the royal children, and suite, along the Bridgewater Canal, to the landing place and pavilion prepared for her debarkation in the grounds of Worsley Hall; whence the royal cortege were conveyed over a new drive, made for the occasion, to the hall, where her majesty rested for the night.'[6]

The boat was specially built by the Earl of Ellesmere, about 45 ft long, with an ornamented bow, central saloon, tiller carved like a snake, and stern post heavily carved with the Ellesmere arms. She was painted white, lined in red and gold, and drawn by two grey horses ridden by postilions. This craft was again used to carry royalty when on 20 July 1867 the Prince and Princess of Wales went from Worsley Hall to Old Trafford to open the Royal Agricultural Show. She then survived as the Earl's private barge.* In 1923 she was fitted with a petrol engine for directors' inspections, and in 1931, the carvings having been removed, became a launch to carry parties about Manchester docks. She was condemned as unsafe in 1936, and sold in 1938 to timber merchants for £5.

Loch was looking for support in the railway world, and chose the Great Western and its allies, whose lines extended past Shrewsbury to Chester, Birkenhead and towards Manchester. The trustees' own waterways and their carrying services to Birmingham and other midland towns could be used to enable Great Western traffic to reach Manchester and Liverpool, while the railways offered them an alternative route for their own midlands-bound goods to those of the Trent & Mersey Canal, controlled by the North Staffordshire Railway, or the Shropshire Union, under the influence of the London & North Western, itself planning to acquire the North Staffordshire. However, by 1855, the year Loch died, it was clear that the Great Western did not intend an alliance, but only to use the trustees to strengthen its own position, and that the trustees could not commit themselves to a railway alliance that fell short of a lease or a sale.

Meanwhile Loch had been having his own troubles with the New Quay company who were independent carriers on the Mersey & Irwell, not covered by an agency agreement. They persisted in giving under-the-counter reductions to their customers which undercut the rates the trustees had agreed with the London & North Western and the Lancashire & Yorkshire, as well as the

* She is illustrated in those days on p. 71 of *The Canal Age*.

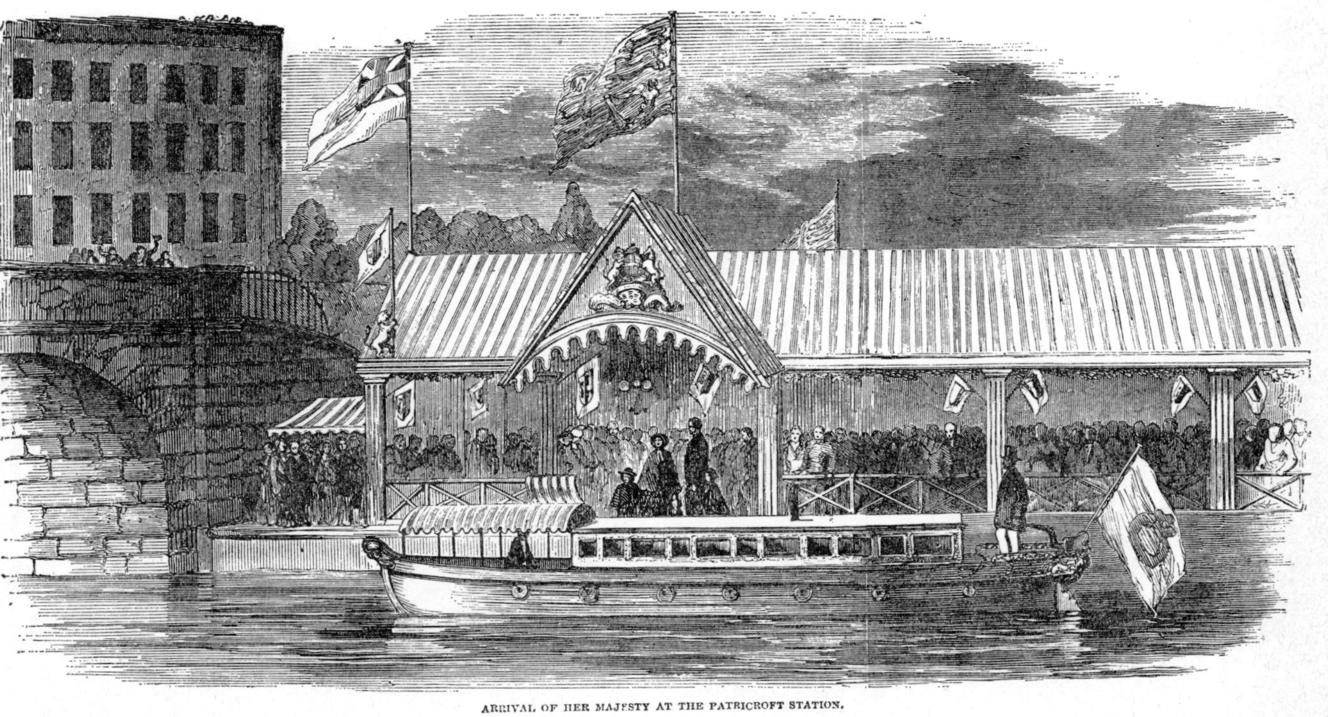
ARRIVAL OF HER MAJESTY AT THE PATRICROFT STATION.

26. Queen Victoria arrives at Patricroft station for her canal trip to Worsley in 1851

other water carriers who were acting as their agents, and Loch found them both influential to harm the trustees and remarkably difficult to deal with.[7]

In December 1852 Fereday Smith, general manager under James Loch as superintendent and his son George as deputy, met the Weaver trustees with a plan of a canal from Runcorn to the basin at Weston Point, mainly to carry salt, and an authorizing Act was passed in 1853.[8] It empowered the Earl of Ellesmere to build the canal at his own expense from the Francis dock at Runcorn, sell it to his trustees, and charge up to 6d a ton. The Weaver trustees might also charge craft that had passed the canal for the use of their Weston Canal (then a free navigation) from the junction to Weston Point.

No work on the canal had, however, been done when Loch died in 1855, to be succeeded as superintendent by Lord Ellesmere's third son, the Hon Algernon Egerton, who remained in charge of the waterways until the Bridgewater Navigation Company took over in 1872. Behind him was Fereday Smith, waiting for his opportunity now Loch was gone. Loch had perhaps been too much of a railway man. In his anxiety to maintain his rates agreements and to ally himself with the Great Western, he had given too little attention to maintaining the real strength of the trustees' business. In the year he died the North Staffordshire Railway, anxious to break out of an encircling ring of railways, and having been prevented, partly by the Bridgewater, from amalgamating with the London & North Western,[9] sought to lease the Bridgewater canals and properties. The railway would have gained an outlet to Manchester, Liverpool and the Mersey, and the trustees the advantage of a fixed income. A Bill was introduced early in 1856, which failed on standing orders, and again in 1857. This was, however, not proceeded with, for by then Fereday Smith had convinced Algernon Egerton and Lord Ellesmere that more could be made of their property than the North Staffordshire had offered.

Fereday Smith had several aims: to improve the organization and efficiency of the trustees' business; to negotiate long-distance toll agreements with neighbouring canal companies, and then to leave carrying to private carriers; and to build the Runcorn & Weston Canal, develop Runcorn docks to forestall ideas of dock building on the Weaver, and improve accommodation at Liverpool and Manchester. He did his best to carry them out in a period of intense railway activity.

Smith did reorganize the administration, and from 1857 onwards began to withdraw from the east midlands and south Staffordshire carrying trades, though not from that on the Weaver. In that year the Earl died, and a new Runcorn & Weston Canal Act[10] had to be passed empowering the trustees to buy the canal from the new Earl and finish it at a cost not to exceed £40,000. The canal, 1¼ miles long, with two locks taking craft 72 ft 3 in. × 18 ft 5 in., was opened late in 1859. It does not seem to have been much used. For instance, out of 36,400 tons of salt from the Weaver for Runcorn docks in 1883, only 4,400 tons went through the canal, the rest going by the estuary. However, it occasionally acted as an alternative entrance to the Bridgewater Canal when access was difficult to Runcorn locks, and then was invaluable.

In 1857 the Mersey Docks & Harbour Board was set up, which bought out the Liverpool town dues for £1½m. The Board was to recoup itself, and shippers expected that before long the old levies would end. The portion attributable to the upper Mersey ports, including Runcorn, was bought out by the Upper Mersey Dues Trustees, to whom the Bridgewater had contributed funds, and of whom Fereday Smith was the first chairman.[11] Delay in finally freeing users of the river from the burden of the repayments was later to be an encouragement to the project of a Ship Canal.

Trade at Runcorn, Potteries traffic, salt transhipment and traffic between the midland canals and Scotland or Ireland, had been increasing for many years, and the trustees now put major improvements in hand. About the beginning of 1860 a large tidal basin was completed there, and in July the Alfred dock, 600 ft × 136 ft, with an entrance 50 ft wide, and equipped with hydraulic cranes. These works, with the Runcorn & Weston Canal, cost some £100,000. The Mersey & Irwell's old basins at Runcorn were not developed; indeed, as much of their transhipment and dock trade as possible had already been transferred to the canal docks. Partly owing to the current congestion at Liverpool, traffic increased at once, and the trustees started on further improvements to enable ships of 750 tons to reach Runcorn on some 300 tides a year. At Liverpool the trustees spent a good deal on improvements at the Duke's dock and setting up a depot at Stanley dock.[12]

In 1862–3 the Wigan coalmasters, anxious to increase sales of coal on the Mersey, and dissatisfied with the Leeds & Liverpool's facilities, discussed with the Bridgewater a scheme to build short mineral railways converging on Leigh, whence a ship canal to take 350 ton craft would run to near Cadishead and then along an enlarged Mersey & Irwell. However, difficulties caused a railway to Runcorn to be proposed instead, and Fereday Smith became a preliminary director of the Lancashire Union Railways, the trustees having agreed to subscribe £7,500 and provide equipment to ship coal.[13]

By 1855 the swift canal packets had ceased to run, and in the following year the boats were sold.[14] In December 1858, ordinary passenger boats were running on the Mersey & Irwell from Warrington bridge to Manchester and back twice a day, (once on Sundays), and to Runcorn once daily. A packet also ran twice a day between Preston Brook station (beside the canal, between the exit from Preston Brook tunnel and the junction with the Bridgewater main line) and Runcorn in conjunction with the trains. From Runcorn the *Countess of Ellesmere* worked daily to Liverpool, and the *Weaver* daily to Northwich.[15] In September 1865 a similar service was still provided on the Mersey & Irwell, but to and from Howley Quay instead of Warrington bridge. But for the Bridgewater Canal the only service was a luggage and passenger packet every weekday between Stockton Quay and Manchester.[16]

The London & North Western Railway's high-level bridge at Runcorn Gap was authorized in 1861 and opened in 1869. On 12 March 1861 the Bridgewater concern made an agreement with the railway, about the bridge and also about railway access to their proposed works at Runcorn. Edward Leader Williams, writing about Runcorn docks in 1865,[17] said that they were crowded, and that the railway company were putting in an expensive branch which would bring them within 13 miles of Winsford. It would be on a high level, and salt would be tipped down chutes at a cost less than carriage on the Weaver. He went on to point out that some of the largest Winsford works already had railway connections, and that others on the Wharton side of the river were having them put in. The salt trade, he added, was rapidly increasing at Runcorn:

Year	*Tons*
1861	112,424
1864	160,548
1865 (9 months)	138,486

In the nine months to 30 September 1865, 2,658 ships arrived at Runcorn. Most were 100–150 ton coasters, the largest being 200–500 tons.

In that year the trustees began work on a second dock at Runcorn as an extension of the Alfred, but did not complete it, partly perhaps because of threatened competition from the Weaver's new

Delamere dock, then being built, partly also because of possible railway purchase. On the side of the Weaver, they moved to avoid a rates war by agreeing in 1869 for similar charges for dockage, use of cranes, and other services at Runcorn or Weston Point. For themselves, they provided more warehouses, hydraulic cranes and dock railways, and busied themselves in seeking new trade.

In 1870 we have a good, though not quite complete, picture of the Bridgewater enterprise shortly before the end of the trustees' management of the waterways.

William Jackson & Sons,
DAILY CARRIERS
TO AND FROM
LIVERPOOL AND MANCHESTER,
ROCHDALE, HEYWOOD, ASHTON, STALYBRIDGE, HYDE, NEWTON, GLOSSOP, HADFIELD, DINTING, MOTTRAM, and all parts of
YORKSHIRE.

WAREHOUSES AND OFFICES:
Duke's Dock, LIVERPOOL,
68 Dale Street, 23 Kennedy Street, and Castlefield, MANCHESTER.

W. J. & SONS beg to observe that their VESSELS are in the best Working Condition, and are surpassed by none for SAFETY REGULARITY and SPEED.

27. A directory advertisement of 1863 for William Jackson & Sons' carrying services

On the Bridgewater Canal itself, net revenue was £58,113, chiefly made up of:

	£
Tolls on independent carriers' trade	20,511
Tolls on agents' trade	10,200
Net revenue of carrying department*	23,114

* Seemingly except the cost of steam towing, £14,214 in 1869.

The carrying department handled 1,090,847 tons, in trustees' boats or in those of the five wholly controlled subsidiaries, the Anderton, Merchants' and Harrington companies, W. Jackson & Sons and Veevers & Co, 184,214 tons of the total being the cross-river trade to and from Birkenhead. Independent carriers took another 1,399,868 tons.

On the Mersey & Irwell revenue was £34,490, of which £26,512 was for tolls:

	£
On independent carriers' trade	5,678
On trustees'* carrying trade	12,813
On agents' trade	7,210
On the Manchester & Salford Junction Canal and the cross-river trade	811

* The Old Quay name was still used, but the business was part of that of the trustees' carrying department.

Tonnages for the independent carriers were 268,871, and the trustees 201,616. In both the above sets of figures the tonnages carried by agents are missing, though the tolls earned are included.

The trustees had not given up hopes of making a favourable railway alliance, and in 1864 had had talks with the North Staffordshire, London & North Western and Lancashire & Yorkshire railways, and seemingly had also contemplated including the Weaver in an arrangement. But the time was not yet ripe.

Eventually, in 1871, a sale was arranged to Sir Edward Watkin and W. P. Price, chairmen respectively of the Manchester, Sheffield & Lincolnshire and the Midland railways. It appears to have arisen out of negotiations by the Cheshire Lines Committee* for land in Manchester and Liverpool belonging to the trustees. To avoid the appearance of direct railway purchase, especially as the Joint Select Committee on Railway and Canal Amalgamations was then sitting, the roundabout method was chosen of the two chairmen themselves buying the property and then forming a new company to take it over.

The purchase was made on 3 July 1872 for £1,115,000, which included the Bridgewater's canals, carrying business and waterway property, and with it ended the waterway concerns of the trustees, though they remained in existence until 1903 to administer other property. The Bridgewater Navigation Company was formed, and the new owners took possession on 1 September 1872, Edward

* The Cheshire Lines Committee was jointly owned by the Manchester, Sheffield & Lincolnshire, Midland and Great Northern Railways.

Leader Williams having agreed to leave the Weaver to become engineer and general manager. To appease Manchester corporation and Chamber of Commerce, who had objected strongly to railway control, the town clerk was given a seat on the new board, and the Chamber was promised that the canals would be fully worked.

Indeed, the new company did behave very independently. On the degree of railway control, there was more than one opinion. W. H. Bartholomew, engineer of the Aire & Calder Navigation, favoured an amalgamation of his company with others, including the Bridgewater, on the east–west route. In 1883 he said that 'the influence of the railways at the board is so great that one can hardly hope . . . that you could successfully arrange an amalgamation',[18] and 'The Bridgewater, practically, are governed by the Cheshire Lines Committee'.[19] On the other hand, James Allport, formerly manager of the Midland Railway, vigorously—perhaps a little too vigorously—maintained on the same occasion that the railway had neither direct nor indirect influence on the Bridgewater board.[20]

When the new company took possession, haulage on the canals and navigations was still by horses, usually two to each barge. Steam traction had been tried earlier, for instance in 1859 with the *Result*, a 16 h.p. tug which could haul five barges on the canal at 3 m.p.h.,[21] but given up. In their prospectus the company had promised that they would try to use steam towing, and at the first shareholders' meeting Watkin said:

> 'Narrow canals, and those with locks or a scarce supply of water, could not possibly compete with railways; but a broad and deep canal like the Bridgewater, if steam could be substituted for horse power, might hold its own.'[22]

Leader Williams first allowed Fowler of Leeds, of steam ploughing fame, to experiment with cable towing at his own expense, a system whereby a cable or chain laid in the waterway's bed was picked up and let fall again as a tug hauled itself along.[23] Fowler claimed that he could reduce towing costs by one-half to two-thirds, but the system was better suited to long hauls on rivers than short on canals, and Leader Williams decided instead to buy narrow-beam tugs. He ordered six in 1875–6, and eventually had some 28. They were used mainly to tow up to three barges on the long lockless run of the canal to Runcorn, where they saved about half the previous cost of horse towing, though they could also work on the Mersey. The money saved was spent on extending the canal walling, a job that was completed in the late 1880s.

One result was to divert most through traffic from the Mersey

& Irwell to the canal. Yet much mainly short haul movement remained. It totalled 676,950 tons in 1884, excluding dredging and maintenance tonnages. The busiest sections were from Runcorn to works along the Latchford canal and to Warrington, and between Throstle Nest and Hunt's Bank at Manchester, much of the latter the result of the 339,950 tons that worked through Hulme locks to and from the canal.

The company's longer range planning, however, looked rather to an enlarged Mersey & Irwell as the main carrier of goods between Manchester and Liverpool: one that would make unnecessary the ship canal that was then beginning to be discussed,[24] which if built would greatly affect both the Bridgewater Navigation Company and the railways. Their first efforts were put into designing a weir to eliminate flooding and prevent the shoaling effect of the fixed weirs. In 1878 a new form of sluice was installed at Howley and Woolston in place of the latter; these caused shoals to be scoured out and increased the depth from 4 ft 6 in. with dredging to 7 ft 6 in. without. A slightly different design of sluice was also put in at Throstle Nest. A cut was made at Mode Wheel to avoid the double bend there, and in 1883 Calamanco (Sandywarps) lock with its 1 ft fall was eliminated by taking out the gates.

The Bridgewater Navigation Company then produced a plan to reduce the number of locks from 10 to 6, increase their ruling dimensions from 76 ft × 15 ft 6 in. to 154 ft × 32 ft (to take four flats at a time or else a steam barge carrying 300–400 tons), and link the Woolston new cut directly with the Runcorn & Latchford Canal to give a single level. These improvements would cut nearly 2 miles of the distance from Albert Bridge, Manchester, to Runcorn, and cost some £325,000.

Giving evidence on the Ship Canal Bill, their representative said:

> 'We think that the Mersey and Irwell navigation, when improved, as we have commenced to improve it, would be sufficient for the conveyance of four or five times the amount of traffic that we now deal with.'

He was supported by W. H. Bartholomew, who considered that after the improvements lightering to Manchester would be about as cheap as ships upon a ship canal, and added that on his navigation 'we find that a canal of this gauge is quite as large as can be utilised, and it can be worked extremely economically'.

When the Bridgewater Navigation Company took over, they thought the traffic offering at Runcorn warranted more accommodation, and restarted work on the trustees' unfinished dock. This, the Fenton dock, named after the company's chairman, was opened on 22 July 1875 at a cost to the company of some £50,000. About 600 ft long and 148 ft wide, it was equipped with hydraulic cranes and tips, railway sidings, and three covered sheds intended mainly for china clay cargoes, which 'are fitted with the most modern improved high-level tramways, on which, the cargoes are conveyed to their respective bays'.[25] In 1875 Leader Williams approached the Weaver to suggest direct communication between Runcorn and Weston Point docks, with the idea that Runcorn traffic would use the Delamere dock entrance. But the trustees, seeing plenty of traffic of their own, refused. In the south-western corner of their Fenton dock the Bridgewater company then built a large lock,* 120 ft × 26 ft, with a centre pair of gates reducing it to 85 ft, and admitting craft drawing 15 ft, which gave access to the first portion of the old Runcorn & Weston Canal, now turned into a ship basin, by widening it for some distance by 14 ft and deepening it to 15 ft. The old canal lock was moved to the end of the widened section nearer Weston Point. Ships lying in what was now called Arnold dock discharged to wharves on the river side. This work was done in 1876; the following year the trade of Runcorn exceeded 500,000 tons, exports being salt, coal and pitch, and imports flints, china clay and other Potteries' raw materials, sulphur, white sand, iron, grain and flour.[26]

The Manchester & Salford Junction Canal still struggled on. In 1869 revenue from tolls and the Irwell trade to and from the Manchester, Bolton & Bury was £935, and in 1870 £811. On 13 April 1872 the Cheshire Lines Committee agreed with the Bridgewater Navigation concern to carry the tracks for their proposed line to Manchester Central over the canal, this undertaking being confirmed in their Act of that year. But another[27] of 1875 empowered them to close and fill in the canal between Lower Mosley Street and Watson Street, on paying compensation to the Bridgewater company which would take account of their release from the 1872 undertaking. The company in turn agreed that the Rochdale company and others who had formerly used the Salford Junction line might freely use the Bridgewater from the Rochdale junction to Hulme locks at a toll not exceeding 3d. When the Great Northern Railway built its goods station in 1899, two hoist wells were dug down to the remaining portion of the canal 25 ft below,

* This lock was later closed. Subsequently the Arnold dock became disused, and is now filled in.

so that traffic might be interchanged between the railway and the ship canal docks by way of the Irwell. This portion of the Manchester & Salford Junction was last regularly used in 1922, and was abandoned under a Ship Canal Act[28] of 1936. The tunnel was used as an air raid shelter during World War II.

In 1876 the Upper Mersey Navigation Act[29] created a commission for lighting and buoying the upper Mersey, from a line drawn from Eastham Ferry to Bank Quay, Warrington, a matter of importance to the waterway undertakings concerned with the river. The commission included representatives of local authorities, trade associations, the Bridgewater Navigation Company, Mersey & Irwell, still nominally in existence, Shropshire Union, London & North Western Railway as owners of the St Helens Canal, and the Weaver trustees. Fereday Smith was one of them.

The Bridgewater company carried now under four names, their own, the Old Quay, the Merchants, and W. Jackson & Sons, on their waterways, the Rochdale, the Ashton and its connections, and the Weaver, and maintained fly-boat services to the Potteries and Birmingham.[30] In 1876, however, they had given up the Anderton company, which was then carried on privately, (though with an agency agreement), and later absorbed the former canal carrying business of the North Staffordshire Railway. On the Liverpool–Manchester run, a sailing table of 1882 shows twice-daily tidal services each way to and from Water Street or Castlefield.[31]

We get a good picture of the Bridgewater and the Mersey & Irwell systems just before their position was radically changed by the Manchester Ship Canal, from the minutes of evidence on the 1885 Ship Canal Bill,[32] and figures in the possession of the Bridgewater Department.

In 1884 total traffic (excluding Runcorn docks) was 2,815,018 tons, of which 1,400,000 tons used the navigations only, and did not pass into the estuary. About 800,000 tons passed over Barton aqueduct, mainly coal from Wigan, Leigh and Worsley to Manchester and beyond. Of the total, about 275,000 tons was exchanged with the Leeds & Liverpool Canal at Leigh, 560,000 tons with the Rochdale, and 480,000 tons with the Trent & Mersey. About 70 per cent of all traffic was carried by carriers who were neither owned by, or agents of, the Bridgewater company. These independents carried 1,752,000 tons an average haul of 16 miles and paid ½d a mile for it, this sum covering toll from Leigh to Runcorn, towage to Liverpool, towage light back to Runcorn, and free passage back empty to Leigh.

The original tolls still applied: 3s 4d (16½p) maximum on the Mersey & Irwell, 2s 6d (12½p) on the canal, with a carrying maximum rate of 6s (30p). This was increased to 8s (40p) for cotton and 9s 2d (46p) for manufactured goods by charges for services such as collection and delivery or cartage in Liverpool, wharfage and storage. Cotton, for instance, could be stored for two months at Manchester if required. The company's average receipts over the whole navigation were 0·515d per ton mile.

Administration was up to date. It was said of through Manchester–Liverpool services in 1885:

> 'We have five or six telegraph stations in Manchester, and as many telephone stations; and our servants wait upon our customers and ascertain for them what vessels are loading and for what ports, and the latest hour goods may be received at Liverpool for particular vessels; they ascertain what bales or cases they have, and have carts waiting their convenience.'

Traffic that could not find a place in the boats went by rail.

In 1885 the Runcorn dock system consisted of two levels of docks connected to the river. On the upper level were the Arnold and Francis docks and the old basin; on the lower the Fenton, Alfred, Tidal and Old docks and a coal basin. A ship lock connected Fenton dock with Arnold on the level above, and two barge locks joined Tidal and Francis. A lock connected Old dock to the coal basin. There were 2½ miles of quay berths, 16 acres of water space, and 37 acres on the quays, with room for expansion. Hydraulic and steam cranes and hoists were installed, and also three hydraulic coal tips. The docks were served by a branch of the London & North Western Railway. For the previous dozen years the average docks trade had been 489,757 tons (281,514 tons being imports), apart from about 900,000 tons a year exchanged between the Bridgewater navigations and Liverpool, which on the river were hauled in trains of barges by steam tugs. In 1884, when total tonnage had been 518,168, salt had accounted for 130,035 tons, pottery clay for 118,102, coal for 71,127, and earthenware exported for 36,759 tons. But clay brought in more revenue than did salt. Most of the salt came off the Trent & Mersey; about a quarter from the Weaver, generally by way of the estuary, hardly at all by the Runcorn & Weston Canal; a little by rail.

In that year also, 55,961 craft passed Runcorn locks. Towage between there and Liverpool was highly organized to suit the tides. Within five hours the Liverpool dock gates would be opened, tugs would collect their trains of barges, deliver them to Runcorn about high water, collect others, take them back to Liverpool and distribute them, and return to their moorings in the Duke's dock before the gates closed. At Runcorn the new flight of locks was used by Liverpool to Manchester traffic, the old line by that going towards Liverpool, which avoided the tidal basin, but had a bottom lock needing 6 ft of water on the sill. At tide time craft leaving the docks were brought to the tidal basin (which was also used as an ordinary ship dock) and the gates closed; the vessels then moved out, ships and barges together, through the one entrance. Barge trains were then made up along the wall running from the tidal entrance to the old line of locks.

At Liverpool in 1885 there was the (Duke's) barge dock of 13 acres in the centre of the dock area, almost the only one not controlled by the Mersey Docks & Harbour Board. The old Mersey & Irwell carriers' dock at the north end of the town, and Stanley dock between that and the Duke's dock, had been let. The Bridgewater trustees' Egerton dock had not been taken over by the Navigation Company, but had instead been transferred to the Board. After the ship canal opened, Duke's dock was sold on 28 April 1900, though a tenancy of part was kept.

Edward Leader Williams left the Bridgewater Navigation Co's service in 1879, being succeeded as general manager by Henry Collier. He first became connected with the ship canal project in 1882, was later appointed engineer, and in 1885 gave evidence before the parliamentary committee against his former employers.

The company did quite well in its later years, paying 8 per cent on its ordinary shares. Receipts and expenditure were:

Year	*Receipts*	*Expenditure*
	£	£
1882	336,018	270,575
1883	327,796	269,323
1884	309,364	251,866
1885	292,708	233,620
1886	280,671	222,123

The story of how the ship canal was promoted has been fully told in Sir Bosdin Leech's great work. The project was naturally opposed by the Navigation Company and, as we have seen, they put forward an alternative though much smaller scheme. It was also bitterly attacked by railway interests, who were happy enough with the existing company, but were likely to lose heavily were the ship canal to be built. After two Bills had been lost, that of 1885[33]

Page 369 Huddersfield Canal: (*above*) the eastern end of Standedge tunnel in 1953; (*below*) Marsden depot in 1933

Page 370 Carlisle Canal at Port Carlisle: (*above*) the sea lock. The remains of the wooden passenger jetty can be seen on the left, and on the right the end of the wharf; (*below*) the former Solway Hotel, built to accommodate passengers changing from canal packet boats to steamers

passed. It provided for the purchase of the Bridgewater company's waterways for £1,710,000 free of debt by the Ship Canal Company—it was of course the Mersey & Irwell they chiefly wanted. Its course between the proposed docks at Manchester and Hunt's Bank was to be maintained, including the approaches to Hulme locks, the Manchester, Bolton & Bury and the Manchester & Salford Junction, but only portions of the rest.

When the ship canal Bill passed, £20,000 was paid on account to the Bridgewater concern but nearly eighteen months later it was reported that nothing had been heard of the Ship Canal Company, though they had to exercise their powers within two years. For the latter it was a race against time to raise capital, but they succeeded, and the purchase cheque was paid over on 4 July 1887, almost the last day.

The ship canal, finished at the end of 1893 and ceremonially opened on 1 January 1894, was built largely on the site of the Mersey & Irwell. Runcorn basin disappeared, and some of the lower part of the Runcorn & Latchford Canal, the rest becoming disused to the point where the ship canal first crossed it near Warrington. Here access to mills and works at Bank Quay and above could be had by turning off the ship canal into the Walton lock branch, 3½ furlongs long. Passing a junction, with on the right a section of the old river used as a ⅞ mile long timber float, craft could pass through the large Walton lock into the Mersey, and so downwards to Bank Quay or upwards past Warrington lock and Howley Quay to the Woolston cut. Alternatively, craft from the ship canal could enter the 1¼ mile final section of the old Runcorn & Latchford cut at Wilderspool (Twenty Steps lock), and at the other end pass through Manor (formerly Latchford) lock into the Mersey, where it could turn left for Howley Quay or right for the Woolston cut. This could be entered at Paddington lock and left at Woolston lock, and so the craft could pass through the Butchersfield cut into the Ship Canal again on the level at Rixton junction.

It was necessary to eliminate Brindley's Barton aqueduct, and to carry the Bridgewater Canal between Worsley and Stretford over the ship canal. The 1885 Bill had proposed a high-level aqueduct with hydraulic lifts at each end. The Bridgewater Company had opposed this, saying that a boat took from 13 to 25 minutes to pass one similar lift at Anderton on the Weaver, too long for the busy Bridgewater. Subsequently Leader Williams designed and built the unique Barton swing aqueduct to carry the canal over the new cut without alteration of level. Hydraulically powered, the swinging

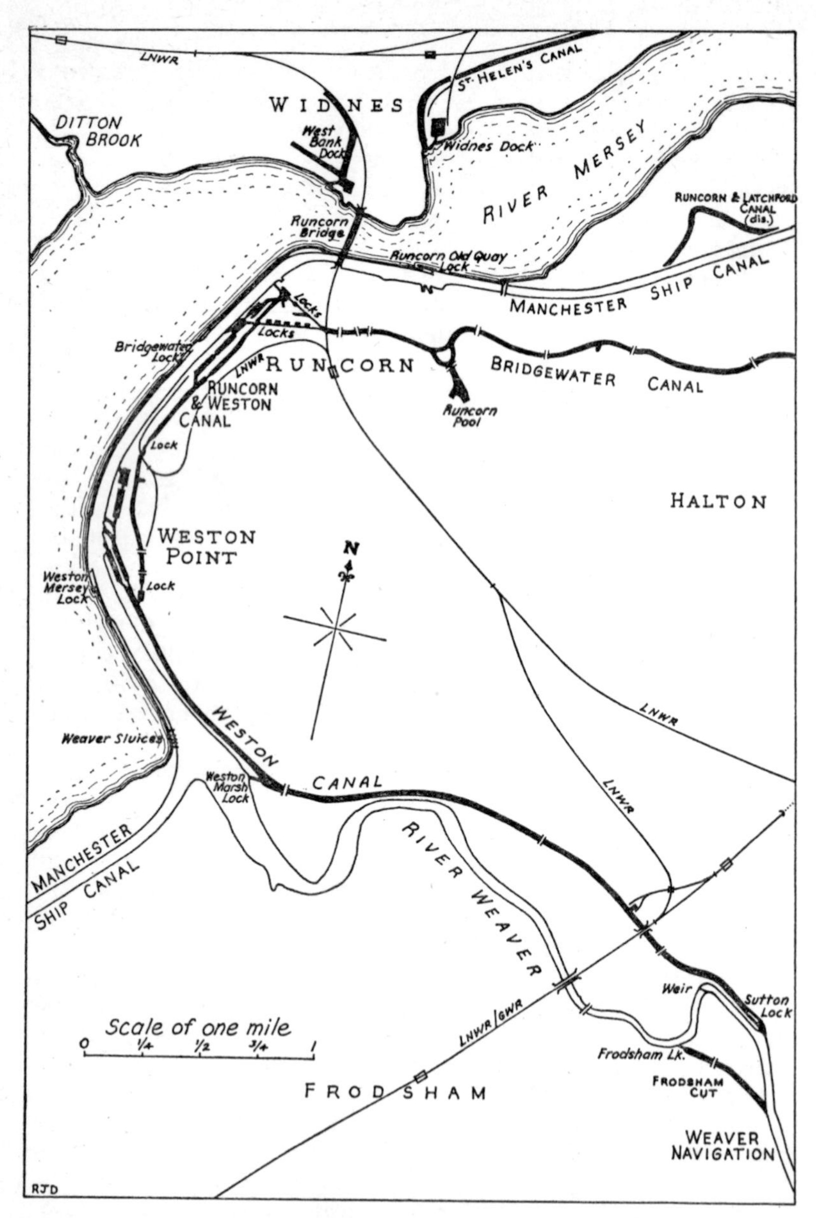

28. The effect of the building of the Manchester Ship Canal upon Runcorn, Weston Point and the Weaver

span is 235 ft long and weighs 1,450 tons. On 29 May 1893, when the new structure was to have been inaugurated, a burst occurred on the Barton approach embankment. However, after repairs, the first barge passed over the swing aqueduct on 21 August, and Brindley's structure was then demolished. The building of the ship canal between the bottom of Runcorn locks and the Mersey also caused a twelve months' closure of Runcorn docks: they were reopened on 9 June 1893. The old lock flight now descended straight into the canal.

The opening of the ship canal naturally caused a fall in traffic that had formerly passed on the navigations to and from Manchester, but did not affect that moving to intermediate points or passing to the Rochdale or the Leeds & Liverpool canals. However, much the greater part of the ship canal's traffic, even in its earliest years, was new, and not transferred, as these figures comparing 1896 with 1900 show:

Commodity	*Liverpool trade* *Increase on ship canal* *tons*	*Decrease on Bridgewater Canal* *tons*
Timber	116,587	17,463
Grain & Flour	152,005	22,851
Cotton	79,112	17,040
Oil	109,241	4,953
Textiles	25,467	10,631
	482,412	72,938

In 1900 the chairman of the Ship Canal Company wrote: 'while the Bridgewater Department has been losing traffic to and from Liverpool which paid well, it has been gradually becoming (we may say rapidly since 1896) more and more an essential valuable adjunct to the Ship Canal on terms which give very little profit to itself but give good toll revenue to the Ship Canal Department', for instance, in providing cartage services, or transhipment to barges at Manchester docks for the upper Irwell, the Rochdale Canal or the Leigh branch.

Traffic was well maintained up to World War I, with tonnages approaching 2,000,000 annually, and the canal itself was steadily improved and widened to a uniform 45 ft. An echo of former days was the *Duchess-Countess*, in August 1913 still running between Knott Mill and Stockton as a parcels and cargo boat, but carrying an occasional passenger by permission. She lasted to about 1924. Business fell off heavily after the war, as less trade took place with connecting waterways, internal movements transferred to road, and tugging services on the Mersey were reduced. These figures summarize the story:

Year	*Total tonnage*	*Bridgewater Department*	*Bye-traders*	*Total through Runcorn locks/R & W Canal*
1900	1,748,869	482,236	1,266,633	585,543
1913	1,811,447	659,404	1,152,043	751,744
1923	1,108,124	305,876	802,248	233,368
1938	815,391	76,342	739,049	200,275
1958	604,884	111,180	493,704	5,378
1962	366,629	116,436	250,193	1,750
1965	256,091	134,091	122,000	–

By 1960 toll revenue of the Bridgewater Department was down to £20,457 and by 1964 to £13,774. More money came from other sources, however, among them the sale of industrial water and pleasure-boat licences and mooring fees. Pleasure craft were allowed to move freely from the early 1950s, and are now encouraged.

In 1947 the Hulme–Stretford length of canal was deepened, a major work, and in 1962 the three locks at Hulme joining the canal to the Irwell were rebuilt as a single lock, the two upper ones being eliminated and the lowest increased in height.

Traffic still passes through Walton lock on the ship canal to Bank Quay and Warrington, and through Twenty Steps and Manor locks to Howley Quay.

From the turn of the century Runcorn docks, already affected by the ship canal, suffered more and more from the general decline of trade upon the canals with which they were connected. Without road access, and in competition with Weston Point and Ellesmere Port for the Potteries traffic, tonnages fell steadily away until the outbreak of World War I, then dropped suddenly and failed to recover. A total of 373,315 tons handled (239,088 being imports) in 1900 had fallen to 124,428 (84,712) in 1925. Apart from Potteries traffic, the main coastwise imports in 1900 were road materials, slates, pig iron and scrap, in that order; in 1926 slates, road materials, aluminium and scrap. Foreign imports of grain and timber were considerable in 1900, but both traffics ceased after 1904 as sailing ships discharging to barges at Runcorn were replaced by steamers going up to Manchester on the ship canal.

After World War II, road and rail transport from Cornwall diverted most of the Potteries traffic from coasting vessels, and by 1957 the dock trade had fallen below 50,000 tons. Then road access was improved and the position bettered. By that time there had been no considerable commercial use either of Runcorn locks or of the Runcorn & Weston Canal* for some years, nor was there traffic to works on the canal at Runcorn itself.

In the late 1950s, encouraged by the building of the Runcorn–Widnes bridge which offered the docks a new outlet, decisions were taken which were radically to change the character of Runcorn docks by divorcing them completely from the canal that gave them birth. These were to close the locks to the ship canal, build a new lock between the ship canal and the docks, close and fill in the Runcorn & Weston Canal, demolish old warehouses built for the Potteries trade and no longer needed, and lay out the area afresh to enable the docks to be as efficiently worked as possible. The last commercial traffic passed down the old flight of locks in 1939 or 1940, along the Runcorn & Weston Canal in 1962, and down the new locks in 1963.

The following figures illustrate the results of the change:

Runcorn docks

Year	*Total trade* tons	*Imported pottery materials* tons	*Sent forward by canal using Runcorn locks or the Runcorn & Weston Canal* tons
1925	124,428	48,785	48,785
1938	140,739	68,211	46,623
1951	32,881	14,116	8,170
1958	52,831	23,592	3,377
1962	171,079	22,970	805
1963	197,273	23,433	
1965	325,046	17,494	

In 1968 the total dock trade reached 435,318 tons, and in 1969 new terminal facilities at Runcorn at a cost of £350,000 were announced. These included enlarging the entrance to the docks from the ship canal by taking 100 yd off the South Pier, demolishing old buildings to increase quay space, and building a new warehouse and transit shed.

* The 1954 figure for the canal was 3,167 tons.

CHAPTER XV

The Weaver and the St Helens

Weaver Navigation

No railway company offered to buy the publicly owned Weaver, but competing lines appeared, there were applications to cross it, notably by the Dutton viaduct, and, later, branches were built into the salt district. One of these, to Winnington, had an exchange wharf with the river that was used to some extent in the 1870s. Railway companies were not anxious to provoke the considerable parliamentary power of the trustees; therefore railway branches into Weaver territory were later and fewer than might have been the case, and protective clauses could usually be agreed in railway Bills. Price-cutting, either against railways or against competing canals, was never an important factor in Weaver history. The trustees tried to keep charges reasonably low and as steady as possible, while following a policy of continuous modernization and improvement of their navigation. It paid them.

The Trent & Mersey Canal had been bought by the North Staffordshire Railway in 1846.[1] That company worked it well, but the Weaver trustees were sensitive to any threat of changed control. In 1854, for instance, they strongly opposed a Bill to amalgamate the North Staffordshire and the London & North Western Railway, minuting that 'the Revenues arising to our Navigation from the Pottery Trade might in a great measure be lost to us'[2] if the London & North Western Railway should take the Anderton transhipment business. Later they were again made nervous by the flirtations of the Bridgewater trustees with various railway companies, and later again by the suggested amalgamation of the North Staffordshire and the Manchester, Sheffield & Lincolnshire Railways in 1876.

Cuts were made to straighten the river, and lock improvement went on, either by rebuilding to larger dimensions, as at Hunts, or by doubling, as at Saltersford in 1848–9 and Pickerings in 1849–50.

Where this was done, new locks 100 ft × 22 ft, with 10 ft over the sills, were placed alongside the old 88 ft × 18 ft locks, with 7 ft 6 in. depth, and enabled much larger craft to pass. In the fifties, as the salt trade of the river above Northwich grew, the trustees decided to double all the upper locks, and to widen this section where they could. In 1857 they wrote of 'the confusion and delay near Winsford owing to the number of vessels loading and discharging at the Straiths or Wharves by which the various works communicate with the Navigation are very serious, and much need some amelioration'.[3] By July 1857 double locks were working at Pickerings, Acton, Saltersford and Winnington. Hartford and Newbridge followed early in the next decade.

It was also necessary to counteract the effects of continual subsidence, of locks, weirs, towpaths and bridges over the river, as a result of rock salt mining and brine pumping. Cubitt in July 1857 referred to the 'extensive and remarkable subsidences which are occurring', and in the same year the trustees wrote: 'At the Junction of the Witton Brook with the River Weaver all semblance of a River has ceased and a Lake of Considerable and rapidly increasing extent now exists.'[4] Because of the effects of subsidence, and also to improve the navigation, the trustees decided in 1859 to remove Northwich lock and weir altogether, and to raise Winnington weir instead.

By 1850 the river had become so prosperous that Weston Point basin became overloaded, the trustees minuting that 'additional accommodation should be afforded at Weston Point for the Trade on the River by the making of an additional Dock and a separate entrance into the Mersey'.[5] Saxon, their engineer, prepared plans, Cubitt approved them, and work was authorized on the New Basin, to the south of the old one. With its entrance lock, it was completed in 1856 at a cost of some £44,000. By 1862 the river was busy enough for night working to start at the locks, and for extra men to be employed at some to help work the boats through quickly.

Saxon resigned as engineer at the end of 1856, having served the river since 1839, and an important appointment was made, that of Edward Leader Williams out of 110 applicants. The son had learned his profession under his father of the same name, engineer of the Severn Commission, and seems quickly to have gained the confidence of the trustees.

From the middle third of the century onwards, we may notice the steady building of chemical works on the river and along the

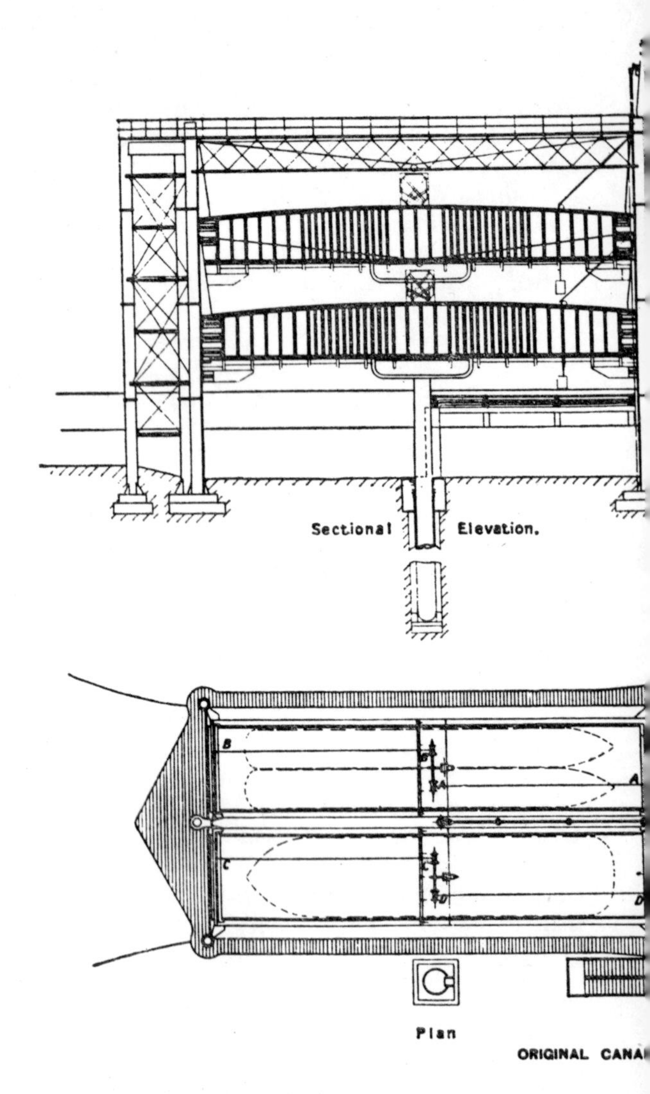

29. Elevation, section and p

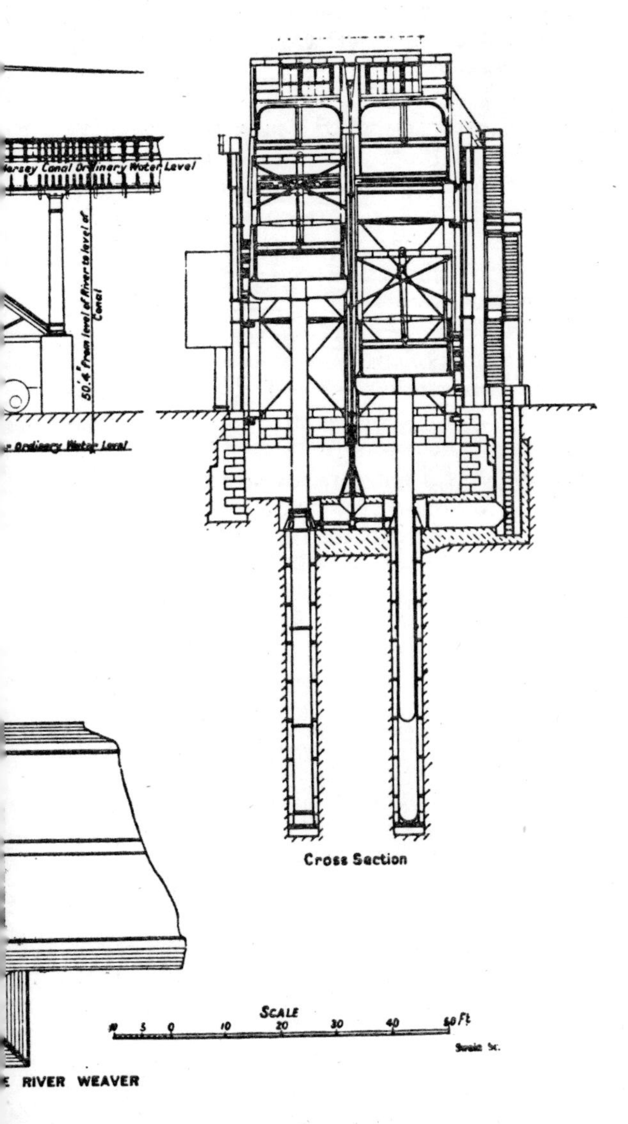

derton lift as originally built

Weston Canal. These were welcome as providing traffic, but they presented problems of water pollution with acids and other troublesome substances, though the trustees did their best to prevent it.

In January 1863 the trustees inaugurated a tugging service for Weaver craft between Weston Point and Liverpool. A steamer appeared on the river itself in 1864, and encouraged them to ask Leader Williams to prepare a report on further improvements, on the one hand taking into account the competition in salt-handling expected from the new dock with its prospective railway branch that the Bridgewater authorities were building at Runcorn, and on the other looking to the extra trade steamers might bring, especially if coasters could also be accommodated.

In 1863 also, the trustees had put in some useful public relations work by giving the Salt Chamber of Commerce, with whom relations had become strained, an excursion trip down the navigation, and then by asking its members for suggestions. As a result, Leader Williams was also told to prepare a plan 'of a Sea Wall at Weston Point to extend from our present Basin to the Bridgewater Property on the Runcorn Side with an entrance Lock and additional Docks and Basins for the purpose of affording further accommodation and facilities to the Trade of the River'.[6] Thereby existing traffic would be given a new deep water entrance, and an added incentive to use Weston Point instead of Runcorn docks, while new business would be attracted. He did so, and was told to go ahead with what was to become the Delamere dock, with a 50 ft wide entrance able to take large coasters. So far the trustees had maintained and improved a navigation that had become steadily more prosperous, and yet in each of the twenty years since 1846–7 had been able to transfer an average of £16,975 p.a. to the county treasurer. Now they had made a decision that was to lead to major changes in their navigation.

Late in 1865 Leader Williams made his recommendations for river improvement. He reported[7] that it was indeed practicable to make the river navigable for seagoing vessels, for which the Delamere dock, then being built, would be suitable. His proposals were for a minimum river depth of 12 ft, with 15 ft lock sills; a new canal and lock connection from Delamere dock to the Old Basin and so to the Weston Canal, which must be enlarged; a new lock to replace the present single one at Sutton, much complained of as being the smallest on the river, though only used at high tides and floods; new locks 200 ft × 40 ft with three pairs of gates, about four times the existing size, to take a steam flat and three lighters totalling 1,000 tons; hydraulic power for lock gates, sluices and swing-bridges (which should be full-width); and movable caps to the weirs. His estimate was £195,800 for the river up as far as Northwich, with a saving if it should prove practicable to remove Winnington lock and get 4 ft more up to Saltersford, or £95,390 if the locks were to be a smaller 120 ft × 23 ft. Steamers, he pointed out, had already reduced delays. A steam flat and three lighters, carrying 800 tons in all and working together through one of his large locks, could reduce the freight (excluding toll) cost of carrying salt from Northwich from 1s 10d (9p) by sailing flat to 9d a ton. Up river, he added, Witton Brook was a natural 80 acre dock for Northwich, while at Anderton, where the trustees owned the land between the river and the Trent & Mersey Canal, 'tramways worked by steam power would enable a large traffic to be transferred . . . even if a direct water communication is not made'.

Leader Williams's proposals were approved by Sir John Hawkshaw, but the trustees decided to deal first with improvements to the lower section of their navigation. In 1866 an Act authorized additional borrowing powers for work on Delamere dock and for widening and improving the Weston Canal and rebuilding Sutton lock to take steamers. It also permitted tolls to be levied for traffic along the Weston Canal, which could no longer be regarded merely as an access channel to the river because of the growth of factories on its banks, and authorized charges for the use of Weston Point and its facilities, and for towage services on the river. By 1870 the Delamere dock was open and the other authorized work had been done at a cost of £60,000; the new charges could therefore be levied. In 1870, also, a coasting steamer came up river to load coal and salt at Anderton for Dublin.

From mid-1868 the tonnage carried began to rise steeply as a result of the improvements, to settle down at about 250,000 tons a year higher than before. In 1870–1 the salt trade for the first time exceeded 1,000,000 tons, and now one begins to see the names of salt companies doing their own carrying replacing those of individuals in the trustees' books: in mid-1860, for instance, the British, Cheshire Amalgamated and Winnington Salt companies, and from 1888 the powerful Salt Union.

Here are figures for the period since 1845:

Years from April	*Toll revenue* £	*Salt tons*	*Coal tons*	*Other tons*	*Total*† *tons*
1846–48	41,594	637,511	152,025	64,936	854,472
1849–51	42,570	652,757	161,427	58,523	872,707
1852–54	42,787	659,483	162,638	51,081	873,202
1855–57	49,464	789,110	201,837	51,106	1,042,053
1858–60	43,777	779,609	190,591	53,621	1,023,821
1861–63	43,930*	782,878	200,967	43,108	1,026,953
1864–66	43,738*	760,337	208,275	58,373	1,026,985
1867–69	54,317*	949,090	258,531	69,046	1,276,667
1870–72	60,504*	1,038,093	248,144	75,901	1,362,138
1873–75	60,533*	1,042,096	192,313	73,528	1,307,937
1876–78	61,197*	1,052,138	174,498	72,959	1,299,595

* These figures from 1863–4 onwards are gross, and contain sums of around £2,500 p.a. which were returned as drawbacks on certain cargoes.

† These tonnage figures exclude those on the Weston Canal after a separate toll began to be levied under the Act of 1866. About 175,000 tons a year were recorded against the Canal, a good proportion of the traffic being to and from local works, and so not passing to the river itself, the rest being counted again in the river figures. On the other hand, the toll revenue figures given include Canal receipts.

Within the figures, salt carryings show two great changes. In 1846–7, 331,137 tons were shipped through Northwich toll-office, 235,908 through Winsford; in 1872–3, Northwich had increased only to 374,443, but Winsford to 639,054, which suggests that much of Northwich's output was going down the Trent & Mersey. Again, in 1846–7 rock salt tonnage was 99,193 out of 567,045; in 1872–3 it was 95,429 out of 1,013,573.

For a long time transhipment at Anderton had gone steadily on. As we have seen, Leader Williams in 1865 had suggested steam-powered tramways should be used to exchange traffic, and had hinted at a water connection. Probably as a result of the Weston improvements, salt traffics transferred there began to increase at the end of 1867, and he was told to re-lay the old tramway, and build separate sets of trucks for coal and salt. Further loading stages were added in 1869. Competition from the Weaver seems sufficiently to have affected the Bridgewater to make them in 1870 suggest an agreement to avoid competition by equalizing charges from Anderton by either route, but the trustees were naturally not tempted. Next came a deputation from the Potteries Chamber of Commerce, and representatives of the north Staffordshire iron trade, wanting to use the Weaver and to be accommodated at Anderton, its leaders offering to guarantee 30,000 tons of potters' materials and 15,000 tons of manufactured goods a year, a considerable increase on the figures of 17,000 and 16,000 tons on which the Bridgewater Navigation Co was then getting its own drawback from the trustees. The Bridgewater now approached the salt trade, offering reductions in its salt tolls in exchange for guaranteed tonnages from the producers, and were told by the Weaver: 'let it be widely known that in the event of such reductions being made, . . . we shall be prepared to take steps to meet the same'.[8]

More important, the trustees now started to think about making a physical junction with the Trent & Mersey at Anderton in order to get a bigger share of traffic from the Bridgewater.[9] A hydraulic lift was mentioned as a possible improvement in the 1866 Act. Now, in July 1870, Leader Williams explained to a sub-committee how a lift would work, that did not cost more than £12,000. Talks were opened with the North Staffordshire Railway, who were shown a model of the proposed lift. But they refused to agree, and compelled the trustees to go ahead on their own. The Staffordshire interests went on pressing, however, and themselves formed a carrying company, the Traders (North Staffs) Carrying Co, to operate via Anderton. So in October 1871 the trustees agreed to raise money not only to continue widening and deepening the river, but also to build a lift and basins. The necessary Acts were passed in 1872, Emmerson & Co were given the lift contract, and Leader Williams got drawings and specifications approved in September. But by then, to the trustees' great regret, he had left to be manager and engineer to their rivals the recently formed Bridgewater Navigation Co, and had been succeeded by J. Watt Sandeman. Changes were made in the design of the lift while building was going on, mainly to strengthen it, and the estimate, itself much higher than Leader Williams's first ideas, was considerably exceeded. Traffic began on 26 July 1875, by which time it had cost £29,463, with another £18,965 for the foundations, set in ground exposed to subsidence.

A number of experimental canal lifts had been built in Britain, at Ruabon on the Ellesmere Canal,[10] Tardebigge on the Worcester & Birmingham,[11] Camden Town on the Regent's[12] and Mells on the Dorset & Somerset.[13] But the set of seven on the Grand Western Canal, worked from 1838 for nearly thirty years, were the only ones successful under operating conditions in this or, as far as we know, in any country until that at Anderton was built.[14]

The idea of a lift was Leader Williams's. The designer was Edwin Clark of Clark, Standfield & Clark, consulting engineers, of Westminster. As built, it had a lift of 50 ft 4 in., using two wrought-iron caissons 75 ft × 15½ ft × 5 ft, rising and falling side by side within a substantial iron framework. Each caisson could take two narrow boats, or one barge—locally called a duker, because it had originated on the Bridgewater Canal—carrying 80–100 tons. Each of these, weighing 240 tons with its water, was supported by a 3 ft diameter iron ram moving vertically in a cast-iron hydraulic press, the two presses being connected by a 5 in. diameter pipe. The lift was normally worked by removing 6 in. of water, or 15 tons, from the bottom caisson. This served to cause the upper one to fall until it was partially immersed in the water at the bottom, its rate of descent being controlled by the speed with which the hydraulic fluid could be transferred from one press to another through the 5 in. pipe. The final lift of 4 ft was obtained by closing the pipe between the presses, and connecting the upper one to a hydraulic accumulator whose power had been stored by a steam engine. The rising caisson was then stopped 6 in. below upper canal level, so that enough water could be run in to replace what had been withdrawn at the bottom.[15] The caissons could also be worked separately, but this took much longer. The top of the lift was connected to the land, and so to a basin leading to the Trent & Mersey Canal, by a wrought-iron aqueduct 162 ft 6 in. long and 34 ft 4 in. wide, built in three spans.

In 1882, one of the presses burst, and its caisson, at the top of the lift with a barge in it, fell to the bottom. However, it fell slowly, and no great damage was done, though it must have given the bargemen a nasty moment. After discussions between L. B. Wells, now engineer, Leader Williams and Emmerson's, thicker cast-iron presses, slightly modified in form, were fitted, and the lift, which had been out of action since late May, was reopened on 4 October. The Anderton design was a good one, and formed the basis of that for the bigger lifts later built at Les Fontinettes in France and on the Canal du Centre in Belgium. On the other hand, its construction seems to have caused no great shift of trade to the Anderton route: it merely made it easier to handle. Much salt, however, still continued to be transhipped by the chutes.

The improvements carried out on the river before the 1872 Act had reduced the number of locks from eleven, some of them single, to nine pairs, Northwich and Winnington having been eliminated. In 1871 the trustees decided to go ahead with reconstruction along the lines Leader Williams had recommended in 1865, and got the salt trade's agreement to their proposals. Over the following twenty years the navigation was completely rebuilt at a cost of over

£200,000. New locks, 220 ft × 42 ft 6 in., with 15 ft of water over the sills, were built at Dutton, Saltersford, Hunts and Vale Royal to make pairs with the larger of the earlier pairs; these four, in place of the navigation's original eleven, now controlled the river down to Frodsham on the old entrance, or Sutton on the Weston Canal, where one new 42 ft wide lock had been added beside the old 14 ft one. The new large locks could take a 300 ton steam flat towing three barges, and carrying 1,000 tons in all, at one locking. The gates were operated by water power, using two Pelton wheels to each gate. Compared with 1878, the time taken from Northwich to Weston had been reduced from 6 hours 24 minutes to 4 hours 38 minutes by 1885. At Weston Point, also, another dock, the Tollemache, had been opened in 1885 between the old basin and the Delamere.

The river was now carrying 1,000,000 tons of salt a year, and another 300,000 tons of china clay, stone flints, bones and bone-ash for the Potteries, and cargoes to and from the chemical works on its banks and those of the Weston Canal. They were moved in steam or sailing flats carrying 70–320 tons each. In 1888 265 craft (65 of them steamers) traded on the river, and made 150 trips a week, not counting narrow boats going to and from the Anderton lift.

Tolls were low for 20 miles of excellent navigation: 1s (5p) a ton for white salt, 6d (2½p) for coal, 8d for merchants' goods and pottery. No Weston Point dock dues were charged craft using the river, and these were also towed free on the Mersey. In April 1886, also, the passage of the Anderton lift was made free to craft paying Weaver toll, though a charge (it had exceptions) was reimposed as part of the toll settlement agreed about 1894 with the Board of Trade under the Railway & Canal Traffic Act, 1888. Yet the trustees' income was now some £60,000 p.a., about a quarter of which was paid over to the county.[16]

Sandeman started upon the programme of lock and river improvements, including dredging to a uniform 12 ft, but seems not to have had the necessary professional competence. He resigned, being succeeded by L. B. Wells, and some years later by the well-known J. A. Saner, who remained as engineer, and later manager also, until 1934. Flood control was important to a navigation as busy as the Weaver. Wells designed movable weirs, flood gates each 15 ft wide, supported by masonry piers, and raised or lowered by overhead gearing, and these controlled floods at about 6 ft above normal level, against the previous 8–12 ft.

In December 1882 the trustees read a parliamentary notice for the proposed Manchester Ship Canal, its line to pass between Weston Point basins and docks and the Mersey itself. The trustees saw the benefit the ship canal might be to them, but all the same, they had to fight hard to get the necessary protection clauses into the company's Act of 1885. When built, the docks were connected to the ship canal by a lock. On the opposite side of the ship canal, a little nearer Eastham, Weston Mersey lock was built, 600 ft long and 45 ft wide, giving on to a navigable channel leading to the Main Deep of the Mersey. Under the Act, the canal had to be built from Eastham to Weston Point, together with the Mersey access channel and Weston Mersey lock, before any work could be done in front of Weston Point docks. Craft to and from Weston used, and may use, this channel when there is sufficient water; at other times they are entitled, subject to certain conditions, to use the ship canal free of dues on the ships and tolls on the cargoes between Eastham and Weston. In addition, craft within certain limits of size to and from the Weaver carrying salt only may use the ship canal free, and those carrying raw materials for the Potteries at half rates, or in some cases free also.

The River Weaver itself was now led across the ship canal and through sluices into the Mersey. Frodsham lock, on the river below the point where the Weston Canal branched off, was left for local traffic, but a new cut into the old river and so into the ship canal was provided from the Weston Canal nearer the docks by way of Weston Marsh lock, 229 ft × 42 ft 8 in., built at the ship canal's cost, and opened in April 1891. The ship canal greatly benefited the Weaver. Weston Point docks had been built deep enough to take vessels drawing 15 ft of water at any tide, but Saner said in 1888 that there were only 146 tides a year when they could get up the Mersey. Once the canal was opened, they could reach the docks at any time.

The trustees did their best to prevent the building of pipelines to carry brine. In 1873, for instance, such a line to Widnes was proposed, and they were supported in opposition by most of the salt trade, and the Local Boards of Northwich and Witton. This came up again in 1890, when the trustees petitioned against it. In 1880 another to carry brine from Northwich to Weston was proposed. In 1889 Brunner Mond were given permission to place syphons under the river below Hunts to carry brine from Witton to their works at Winnington, on condition that they did not 'convey any Brine out of the District from any of their shafts at Witton, Winnington or Anderton'.[17] It was a losing battle, their greatest set-

Page 387 (*above*) Unloading baled cotton at Church wharf, Accrington, on the Leeds & Liverpool Canal *c.* 1910; (*below*) a coaster alongside the jetty at the entrance to the Ulverston Canal

Page 388 Lancaster Canal: (*above*) loading coal from a railway wagon tippler at Preston basin *c.* 1923; (*below*) the west end of Hincaster tunnel *c.* 1900

back being in 1910, when the Salt Union opened a new works at Weston Point, supplied by brine pipes, which heavily reduced salt carrying on the river.

From the early 1880s, the river started seriously to lose its salt trade to pipeline and rail: salt tonnage, 1,250,543 in 1880–1, was down to 691,132 in 1892–3, while coal had fallen even faster from 193,509 tons to 25,853. But, fortunately for the Weaver, its policy of continuous modernization had attracted chemical works, notably Brunner, Mond & Co, whose successors, Imperial Chemical Industries, provide almost all the river's living today. Therefore the category of 'other goods' shot up from 89,278 tons in 1880–1 to 660,571 tons in 1892–3, which so far compensated for the loss of so much salt and coal carrying that the river's total tonnages only fell from 1,533,330 to 1,351,703, and those on the canal from 218,946 to 161,354.

The County Councils Act of 1888, which replaced the old Quarter Sessions by new elected bodies, made the past status of the trustees obsolete. Clearly changes had to be made in a body which perpetuated itself by co-option, and which was not formally linked to the new Cheshire County Council. The trustees had, however, to some extent anticipated change, when in 1866 they had agreed that representatives from neighbouring towns should be nominated to the Trust as vacancies occurred. Thereafter, when one of these representatives retired or died, they asked the town to submit names from which they could make another nomination.

Reconstruction was envisaged in an Act of 1893, and as an interim move, representatives of the Local Boards of Winsford and Northwich were then added to the Trust. After long negotiations between the county council, traders, towns and trustees, with many proposals and counter-proposals from the county and the traders, each of whom wanted to get control, Parliament in 1895 finally authorized a new body of 38 members, which took office in November. Ten of these would be existing trustees for life, vacancies being filled by the county council, but with five at least who were not council members, though resident in or near the county, and with the present trustees' property qualification; twelve would be members of the Cheshire County Council, appointed by that body; fourteen representatives of the traders, of whom Salt Union Ltd was the biggest, but with not more than seven from any one interest; and one each from Winsford and Northwich. On the financial side, the Act provided that the first £25,000 of surplus income should be put to reserve; the next

£25,000 should be divided equally between the county council and the reserves; and sums over that, one-third to reserves and two-thirds to the county. But in fact after 1890 no further payments were made to the county until 1947, when the reserves stood at £25,106 and £53 was paid over.

From the early 1890s to 1914, the Weaver's trade was good, its prospects hopeful, and its trustees' policy expansionist. The salt trade was tending downwards and that in coal had almost disappeared, but the carrying of raw materials for the Potteries was expanding (though the river was getting a falling share of the crates traffic in finished goods), and so was chemicals business.

Electricity replaced steam at Northwich maintenance yard in 1901, and two years later did so at Anderton lift to power the hydraulic pumps and caisson gates. That done, it was found to the trustees' consternation that the hydraulic rams and cylinders needed renewal, a job likely to cost £20,000 and close the lift for nine months. They therefore did away with the hydraulic rams, and instead built a new framework over the lift to enable each caisson to be operated separately by using counterbalance weights aided by electric power. This was done, with few interruptions to trade, by 29 July 1908, when the official opening took place. The cost was £25,869, the result much quicker operation and lower running costs.[18] In 1913 the lift transferred a record 226,000 tons of traffic, more than it could have handled before reconstruction.

On the river the 1865 programme of lock elimination and rebuilding ended when Newbridge locks were removed in 1897. Meanwhile, under an Act of 1893, two wide-channel electrically powered swing-bridges were put in at Northwich, the second in 1899. These were designed to survive subsidence without damage, for 90 per cent of their weight was carried by a buoyancy tank, which made them, in effect, floating bridges. Two similar but larger swing bridges were constructed later, at Sutton Weaver in 1923 and Acton Bridge in 1932: at the latter 560 tons of the total weight of 650 tons are carried by the buoyancy caisson. Thenceforward only one low-level stone bridge remained to prevent full-sized craft reaching Winsford, that at Hartford above Northwich. But, after all they had spent at Anderton and on their bridges, the trustees postponed its removal; it survived until 1938, when it was replaced by a bridge with 30 ft headroom enabling coasters at last to navigate to and from Winsford.

They strongly supported the 1909 recommendation of the Royal Commission on Canals[19] for the nationalization of the 'Cross'—

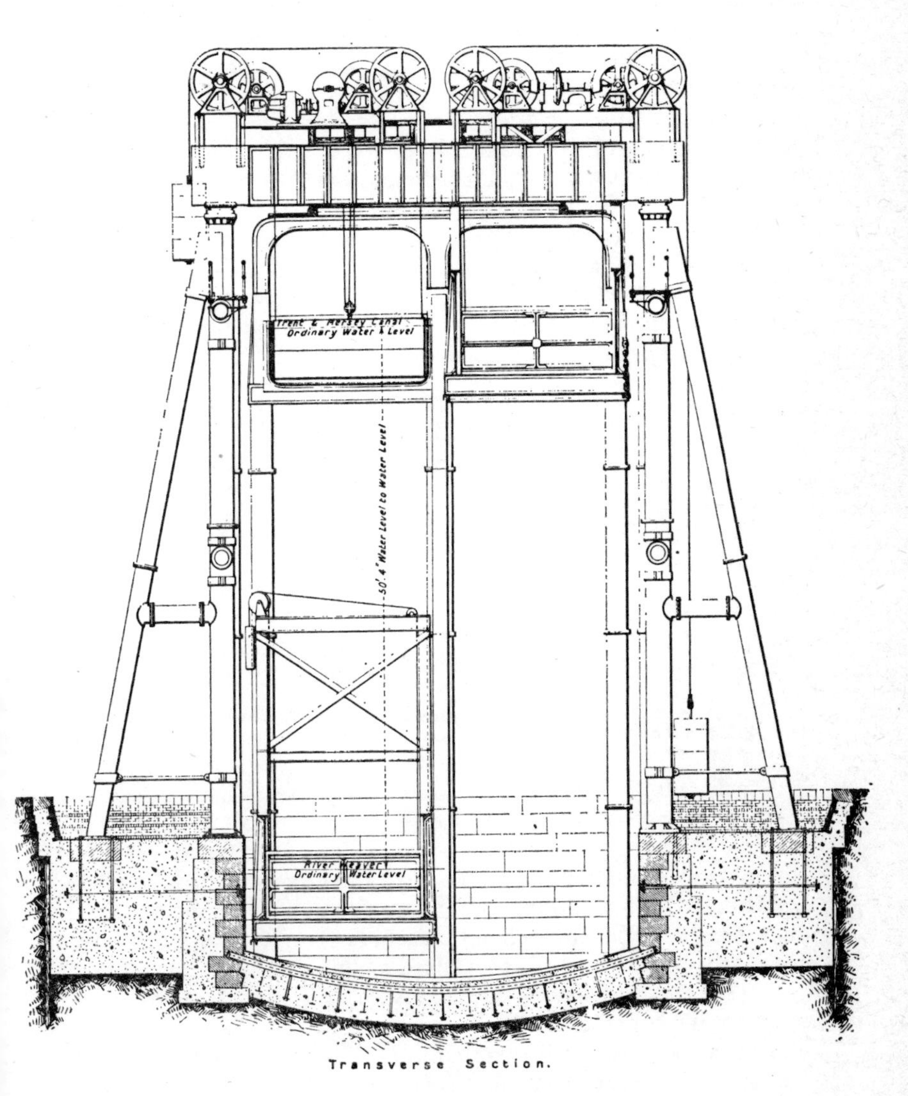

30. J. A. Saner's drawing of the reconstructed Anderton lift

which included their own navigation. Meanwhile, Saner pressed for the Weaver's extension to the Potteries, though he got no further than a conference of local authorities at Stoke-on-Trent in 1912.

Transport routes dislocated by war seldom fall back into their old pattern. World War I did this for the salt trade, as the accompanying figures show. Though chemicals improved and pottery raw materials partially recovered, the inter-war period found the navigation averaging not much over half the tonnages of the century's first decade. Therefore the trustees were cautious and economical, undertaking little new work but continuing to pay off debt incurred in the time of reconstruction until by 1935 they owed nothing.

Years averaged	*Tolls (River & Weston Canal)* £	*Salt (River) tons*	*Pottery materials and crates (River) tons*	*Chemicals (River) tons*	*Total tonnage (River) tons*	*Total River & Weston Canal tons*
1896–7/1900–01	45,788	576,887	71,236	165,307	1,020,424	1,164,265
1901–02/05–06	47,134	558,349	85,641	157,709	1,024,188	1,152,415
1906–07/10–11	49,018	520,600	90,818	212,644	1,012,994	1,132,300
1911–12/15–16	47,793	390,604	90,223	300,198	974,513	1,107,604
1916–17/1920*	34,090	235,232	34,541	240,127	631,201	741,948
1921–25	71,534	185,366	59,224	274,314	586,414	689,852
1926–30	69,207	197,746	59,377	295,039	640,656	757,993
1931–35	58,028	174,684	43,000	273,563	563,348	627,023
1936–40	59,660	139,753	35,373	308,710	562,560	628,910
1941–45	72,609	95,722	11,930	244,634	456,838	532,293

* In 1917 the end of the financial year was changed from 31 March to 31 December. This period is therefore three months short of five years.

A little before World War II a new policy developed, of making small improvements up and down the river that would allow larger coasters to work to Anderton and Northwich, and so eliminate transhipment costs. To 1939 the largest to use the river had been 117 ft long and carried 200 tons, but in 1940 one of 142 ft negotiated it to Northwich and another of 137 ft, the *Jolly Days*, chartered to ICI (Alkali) Ltd, started running regularly.

In 1943 the Ministry of War Transport asked C. M. Marsh, the navigation's engineer and manager, to report on the possibility of a 100 ton waterway to Wolverhampton. As he proposed to utilize the upper Weaver from Winsford to Audlem and thence the Shropshire Union Canal, the trustees secured an Act in 1945 empowering them to extend to Nantwich. Under the same Act Pickerings lock, no longer used, was removed. In 1948 the navigation was transferred to the British Transport Commission, and in

1962 to the British Waterways Board. Frodsham lock, no longer required, was closed in 1955.

World War II again dislocated the salt and the already declining pottery materials trades, leaving that in chemicals as the standby of the post-war river. Since then the navigation has been further improved. The old depot at Anderton has since 1964 been developed for use by coasters, and the river itself deepened to take craft drawing 10 ft 6 in., with the intention of further deepening it over the next few years.

When the war began, Weston Point docks were handling about the same volume of cargo, under 200,000 tons a year, as they had in 1900. By 1948, however, the figure had fallen to a negligible figure, and in 1954 was only 40,000 tons. This rose to 106,000 tons in 1962. Since then the British Waterways Board have made great changes, as can be seen from the plans in Fig 31, adding new sheds, plant, roadways and container berths, and filling in what was no longer needed. In 1969 the tonnage handled reached 451,000.

St Helens Canal

The agreed policy of the new amalgamated company's board was to extend railway branches to the Blackbrook collieries and on towards Wigan, improve the entrance to the dock and canal, seek a junction with the Grand Junction Railway's proposed Runcorn and Huyton line to get direct rail access to the Cheshire salt area, extend a railway along the Mersey towards Liverpool, and join any other railway if it would be beneficial to do so. The company was therefore to be developed on the railway side, while the canal would be kept much as it was. Garston was chosen as the new dock, and the line to it was opened in 1852. A railway running parallel to the canal for some miles was also built in the opposite direction to Warrington and opened in 1853, instead of the canal that the Sankey had had in mind in 1843. At Warrington it joined the Warrington & Stockport Railway, which had the same chairman and secretary as the St Helens company, and which in 1854 was completed to join the Manchester, South Junction & Altrincham at Timperley, to give a continuous line from Manchester to the outskirts of Liverpool, which was of particular interest to the Manchester, Sheffield & Lincolnshire Railway, which had supported the Warrington & Stockport and operated it.

Nevertheless, the canal side of the business was not neglected. In 1849, drawbacks were being given on coal sent up the Shropshire

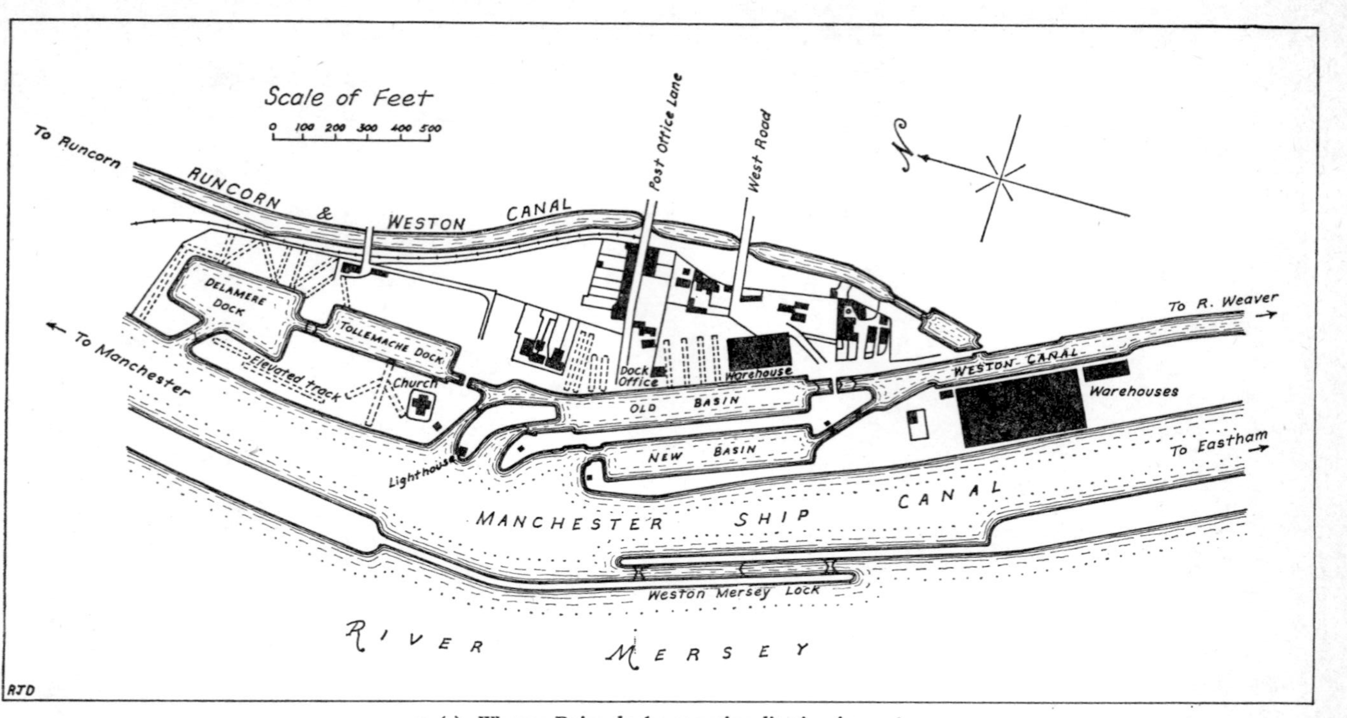

31 (*a*). Weston Point docks at nationalization in 1948

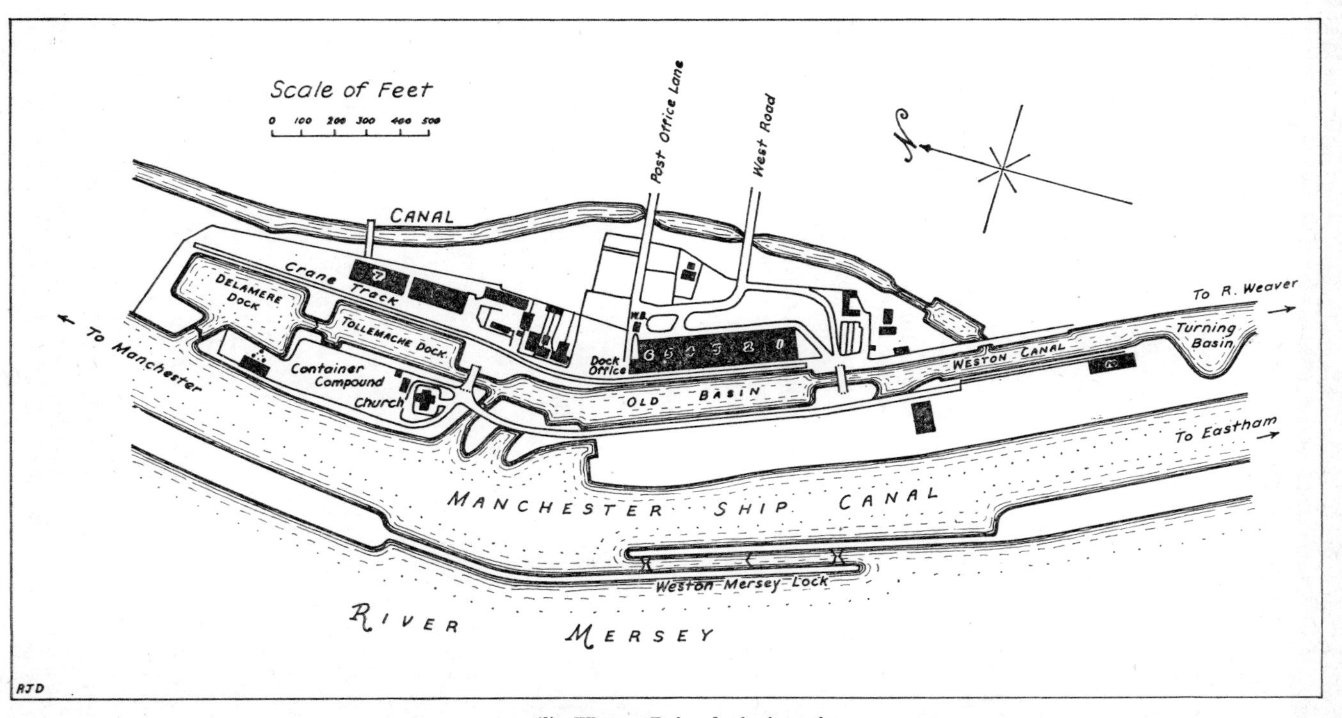

31 (*b*). Weston Point docks in 1969

Union and the Leeds & Liverpool canals as well as the Weaver. In 1851 a 12 h.p. steam engine was ordered to pump water back up the New Double locks at Pocket Nook. The second lock at Fiddlers Ferry was closed in 1846, the 'old canal' being filled in eight years later when the main lock there was repaired. In 1853, as a result of the opening of the railway to Garston and the dock there, the company's railway tonnage first exceeded that of the canal: 613,805 tons by rail and 510,668 by canal. Nevertheless, canal tonnages went on increasing: for the first half of 1856 they were 301,963 tons against 257,396 for the first half of 1853. Sales of industrial water from the canal seem to have begun in 1847. For this reason among others, pollution had to be avoided, and in the 1850s the company had difficulties with Crosfield Bros & Co, who were alleged to be discharging acids into the water.

In 1856 the North Staffordshire Railway unsuccessfully sought running powers over the St Helens line, authority to extend it to Liverpool, and power to buy or amalgamate with the company. This and the Manchester, Sheffield & Lincolnshire Railway's potential rivalry to the London & North Western's Liverpool and Manchester line were naturally unpalatable to the latter. At the beginning of 1859 an agreement covering their many disputes was reached between the London & North Western and the Manchester, Sheffield & Lincolnshire. Very soon the Warrington & Stockport was leased jointly to the St Helens company and the North Western. From 31 October the latter worked trains through to Warrington. Then in 1860 the North Western took over the sole lease of the Warrington–Garston line of the St Helens company, and in 1864 absorbed the whole of the St Helens railway and canal, the old company being dissolved.[20]

Under the Act the 1845 powers to close parts of the canal under certain conditions were replaced by the London and North-Western undertaking to keep it all open, clean, dredged, in repair, and well supplied with water to a depth of 6 ft 3 in., and to afford all facilities for traffic. The exemptions from toll of the original 1755 Act were all continued, and it was enacted that should any of the locks need rebuilding, they should not be made smaller than that existing at Newton Common. Traders at St Helens got themselves very favourable railway rates under the Act (which were continued under that of 1891), between St Helens and Widnes, including dock dues and transfer charges to vessels. Canal tolls were also revised, now giving a maximum of 8d a ton on coal, and 10d on other commodities. A note among the railway papers says that the canal was in 'miserable order', and had to have £23,000 spent on it.[21]

In the seventies the canal is said to have been seriously affected by chemical action upon the mortar of its walling, due to waste from chemical works on its banks. The polluted water also sometimes overflowed into the Sankey Brook, and in times of flood rendered much land barren. The London & North Western obtained an injunction against the manufacturers, and in the result had to buy the meadows as the quickest way of settling the question of compensation. They then rebuilt locks and walls.[22]

Coalmining under the canal caused some minor subsidences, which its railway owners dealt with. But the problem became more serious after 1877—there was one case of a drop of 18 ft in 12 months—and the railway company sued the coalowners, but lost their case in 1892. However, the judgement was reversed in 1893 on appeal.[23] An arrangement was then made that the colliery company should maintain the canal banks, since they knew best when their operations were liable to cause subsidence.[24]

At the end of the century, craft 68 ft × 16 ft 9 in., carrying up to 75 tons, could use the canal, thanks to the depth of 6 ft 3 in., though smaller craft of 55 ft × 14 ft 3 in., carrying 57 tons, worked on it. The Mersey locks at Widnes also admitted lighters working thence to Liverpool, which used only the lowest few hundred yards of the canal and the dock. These were 76 ft × 19 ft 8 in., carried 150 tons, and took advantage of the greater depth of that part of the canal, 7 ft 6 in. in winter and 7 ft in summer. Traffic was alkali, soap, silicate, river sand, acid, sugar, oil, tallow, manure, copper ore, silver sand, salt and copper. There was no longer any coal. Tonnage figures were: 1888, 503,978; 1898, 381,863; 1905, 292,985;* and those for tolls £6,275, £4,235 and £3,010 respectively.

In 1898 about ½ mile of the Ravenhead arm beyond Boardman's Bridge was closed. A report of 8 May 1914 said that 'For several years there has been a considerable decrease in the number of canal boats coming into the borough of St Helens; it is reported that not more than 20 boats carrying cargo have passed through the New Double Locks at Pocket Nook to any of the works on the canal during the last fourteen years'.[25] In 1919 seven flats passed through Newton Common lock, the last to work up to St Helens. In 1920 another stretch of the Ravenhead arm was closed.[26] The Corporation now wanted to widen roads over the canal arms and to remove

* These figures do not include the considerable tonnages which were exempt from toll.

the narrow swing-bridges that created bottlenecks on their roads. An enquiry into the canal was held in November 1930, at the instance of the London, Midland & Scottish Railway company, to whom the canal had been transferred under the Railways Act of 1921, who were losing £1,000 p.a., and of St Helens corporation. This resulted in the closing in 1931 of over 5 miles of the canal north of Newton Common lock at Earlestown,[27] the channel being retained as a water feeder. Below that, diesel barges continued to work to the Sankey Sugar Company's wharf. The closed section included 2⅛ miles of main line, and the Gerard's Bridge, Blackbrook, and Boardman's Bridge branches. Bridges in St Helens were fixed, at Raven Street on 21 July 1932, at Redgate and Old Fold double lock on 17 April 1934, and at Pocket Nook on 22 November 1934.

Traffic and receipts during the present century ran as follows:

Year	*Tonnage* tons	*Tolls* £	*Total Receipts* £	*Expenditur* £
1913	211,167	–	3,777	4,272
1922	135,263	–	5,814	8,911
1930	150,284	–	4,938	6,897
1938	94,016	2,851	4,289	6,660
1946	20,638	826	2,558	9,366

In 1957 about 35,000 tons a year were being carried, raw sugar to Sankey, lead to Sankey Bridges, and chemicals from the Widnes area.

All traffic ceased in 1959 when bulk transport of sugar was introduced, and the canal was abandoned in 1963.

CHAPTER XVI

The Leeds & Liverpool

RAILWAY competition started to bite in 1847, the last year in which the 34 per cent peak dividend was paid; hence the hasty setting up of a carrying department from 1 July 1848. The sub-committee's report on the new establishment concluded:

> 'With regard to the profitable conducting of the carrying business we beg to remind the Committee that this Company has been reluctantly compelled to adopt it, but not with a view to realising any profit from the carrying itself, but simply to maintain the Traffic upon the Canal and thereby secure the Tonnage for the company.'[1]

The railways' reaction was to seek an agreement on rates, which proved impossible, so instead they tried to force the canal company to raise its merchandise tolls to the railways' level in return for a guarantee based on the carrying receipts for the second half of 1849. They probably got wind of the losses on carrying which the company was incurring after a brief initial burst of profitability—£6,000 had been lost against a capital outlay of £15,000 by April 1850—and realized their position was stronger than they thought. The company refused, and dropped its 30 per cent dividend of 1848 to 27 per cent for 1849, followed by a nosedive to 15 per cent in 1850. To reduce expenses, the Bradford buildings were sold and the head office transferred to Liverpool in 1852.

By July 1850 negotiations had been reopened and a draft agreement prepared under which the company undertook to raise tolls on all merchandise and other traffic except coal, bricks, and minerals to 1½d per ton per mile, and lease them to the London & North Western, Midland and Lancashire & Yorkshire railways for 21 years in return for an annual rental of £40,500. The lease was not directly to the railways, but to canal companies they owned, a simple stratagem made possible by the Canal Carriers Act of 1847 which enabled one canal company to lease another. The East

Lancashire Railway, not being a canal owner, was to be made party to the agreement by a deed of arrangement with the lessees.

To counter a lawsuit to prevent the lease by the Lancaster Canal Company who, of course, were interested in the tolls they received from the Leeds & Liverpool over the South End of their canal from Johnson's Hillock to Wigan, a further lease was arranged whereby the Leeds & Liverpool took all tolls on the South End on similar traffic to that leased to the railways, in return for a payment of £4,335 p.a. to the Lancaster for 21 years. At the same time a second lease was entered into with the railways in respect of warehouses, wharves, cranes and other fixtures for £500 p.a., and sale of the stock of the carrying department (horses, boats, etc) for a lump sum of £13,880. Thus the Leeds & Liverpool received from the railways a total annual rental of £41,000, out of which they paid the Lancaster £4,335, leaving a balance of £36,665 a year. The railway leases commenced on 5 August 1850, and the Lancaster lease on 10 March 1851. The railways divided the tolls as follows: London & North Western 10/67ths; Midland 24/67ths; Lancashire & Yorkshire 33/67ths.

Carrying continued as the Leeds & Liverpool Canal Carrying Company. Despite increasing tolls on merchandise by 1s (5p) per ton in 1851, bye-traders continued in business and were carrying at lower rates than the lessees by canal or by rail. The railways' object was to continue carrying profitable goods themselves on the canal, but to drive off the bye-traders and send other goods by rail. Hired boats were used to a considerable extent, it being more economical to use them for one-way traffic such as grain from Liverpool, the owner being able to return with a load of minerals from which the railways were precluded by the terms of the lease. Consequently the lessees found they could not entirely drive the bye-traders from the canal without losing valuable traffic themselves; also the independents continued to flourish on shorter intermediate voyages which the lessees found uneconomic.[2]

The Midland Railway apparently thought better of its participation and in 1857 asked to be allowed to drop out of the consortium at the end of 14 years, but the canal committee took no notice. When the lease expired in 1871, though, they were not party to a two years' extension which the Leeds & Liverpool negotiated with the London & North Western and Lancashire & Yorkshire Railways. In 1872 memorials were presented by Wigan and Burnley corporations, and millowners, merchants and manufacturers in Burnley and Blackburn, urging the company to resume the carriage of merchandise owing to the great delays occasioned during the railways' lease. Probably more important, there was now money in the coffers to finance it. Dividends had climbed thus: 1850, 15 per cent; 1855, 25 per cent; 1860, 27½ per cent; 1865, 28 per cent; 1870, 25½ per cent. Accordingly, on 4 August 1874, the Leeds & Liverpool company once more assumed full control over their revenue and energetically prepared to resume carrying. New warehouses were authorized, existing accommodation repaired or extended all along the line of the canal, and a traffic committee appointed. Advantage was also taken of the newly passed Regulation of Railways Act, 1873, to negotiate through rates with the Aire & Calder, Calder & Hebble and Barnsley canals. On 24 August 1874, twenty days after the end of the lease, the traffic committee reported that the amount of traffic offered was embarrassingly large.

During the lease period, the canal company had not been idle. Under the Lancaster Canal Transfer Act of 1864,[3] the North End of the Lancaster had been leased to the London & North Western Railway and, after representations, the South End between Walton Summit and Wigan Top Lock, including Johnson's Hillock locks, to the Leeds & Liverpool for £7,075 p.a. in perpetuity from 1 August. In 1854 and 1864 there were moves to make a second connection with the docks at Liverpool, but without result.

The vicissitudes of the Bradford Canal caused almost continuous concern from 1865 to 1874 (see p. 412) and in 1854–5 a dispute arose with the Ribble Navigation Company over their Improvement Bill. The Ribble company sought to charge a toll on goods landed in the tidal Douglas below Tarleton lock, which caused the Leeds & Liverpool to commence legal proceedings to restrain them. The dispute ended by the canal company agreeing to pay the river company £50 p.a. for buoying the channel to the mouth of the Douglas for the benefit of canal craft, and maintenance.

In Yorkshire, two rail connections were made, with the Midland Railway where it ran alongside the canal at Niffany's, west of Skipton, by means of a siding and wharf in 1849–50,* and with the Leeds & Thirsk (later Leeds Northern, and finally North Eastern) Railway, which was given permission to construct a wharf near St Annes Ing lock at Leeds in 1849. Renewed railway competition in Lancashire in 1864 arose over the promotion by prominent coalowners, supported by the London & North Western, of the Lancashire Union Railways, designed to run from Blackburn through Chorley to Wigan and St Helens, closely parallel to the

* The site can still be seen.

Leeds & Liverpool and the South End of the Lancaster. It aimed at providing better facilities for the pits on the latter to transport coal to Liverpool. The London & North Western proposed to transfer their St Helens Canal to the new concern, whereby coal traffic could be taken on from St Helens by barge to Garston Docks. This direct attack was strenuously opposed by the Leeds & Liverpool, resulting in the Wigan–St Helens section being deleted from the Bill and the insertion of protective clauses. An Act was obtained in the same year and the line, in two separate sections from Cherry Tree to Chorley and Adlington to Boar's Head, was opened in 1869, under joint London & North Western and Lancashire & Yorkshire ownership.[4]

Between 1855 and 1859 several miles of the Liverpool end of the canal were widened and Shipley wharf was enlarged. In 1863, approval was given to duplicating Ell Meadow lock below Wigan. Steam power appears to have first been used with any regularity in 1866 when the Aire Dale Steam Tug Co were given permission to operate on the river between Leeds Bridge and the entrance lock to the canal, as an extension of their activities on the Aire & Calder, but it was not until 1871 that steamers appeared on the canal proper after further trials with a patent tug by Richard Tennant and Robert Thomas the previous year. As a result four tugs were ordered for the long level pound between Appley lock and Liverpool for the coal traffic, and a steam dredger to deepen the canal wherever possible.

Revenue—1875–1898[5]

Year	*Totals* *Coal* £	*Totals* *Merchandise* £	*Total Revenue*[6] £	*Dividend per cent*
1875	24,533	28,198	70,554	17
1880	–	–	95,521	21
1885	–	–	91,890	17
1889	–	–	88,497	17
1890	22,085	43,409	65,562	17
1898	13,371	21,580	80,150	5½*
1901	–	–	–	1*

* In these years a preference dividend of 3½ per cent was paid.

The period 1874–1901 saw steady expenditure on improving the canal in a successful attempt to maintain traffic in the face of increasing railway competition, but at the cost of reduced rates and, after 1890, a sharp drop in profits and dividends. Although incomplete, the following figures show in particular the big effort made to increase merchandise traffic, to which the rise in average mileage was entirely due. Unfortunately this was the traffic which required the greatest expenditure to foster, by way of enlarged and new warehouse accommodation and so on, with disastrous results on dividends, as the table opposite shows.

Average Carrying Distances—1871–1898

Year	*Coal*	*Merchandise*	*Limestone*	*Manure etc*	*Stone*	*Total*
	Miles	*Miles*	*Miles*	*Miles*	*Miles*	*Miles*
1871	17·3	11·59	19·26	14·75	15·07	16·41
1881	17·62	33·07	15·85	9·68	16·28	20·47
1891	16·89	34·96	18·56	9·1	12·47	20·65
1898	16·96	37·09	18·12	10·2	14·37	21·81

In the first years several new branches were mooted. Accrington Local Board, supported by local tradesmen, asked for one in 1875, and Colne in 1880. In 1882 Accrington promoted a Bill, but was forced to abandon it owing to lack of support from the canal company, but additional accommodation was provided at Liverpool, Blackburn, Leeds and elsewhere from 1876 to 1882, and the canal was widened at Litherland. These years witnessed also a curious development at Tarleton where in 1880 a branch railway was opened from the West Lancashire Railway at Hesketh Bank to a terminus on land alongside the tide lock into the Douglas, which the Leeds & Liverpool leased to the railway for 20s a year. The West Lancashire Railway was bankrupt for most of its existence until taken over by the Lancashire & Yorkshire, and eventually built a line from Preston to Southport. At this time, however, it had only a somewhat useless piece of line from a temporary terminus in Southport to Hesketh Bank, with no connections to any other railways, and it seems that Edward Holden, the chairman and a leading creditor, who appears to have built the branch with possible assistance from the landowner Hesketh, saw it as a way of securing traffic to and from the outside world via the canal, and so gain some revenue. The West Lancashire's Act of 1878 had authorized the railway to 'build, purchase, hire, provide, charter, employ and maintain' steam vessels to carry passengers, livestock and goods from Hesketh-with-Becconsall on the River Douglas to Preston, Lytham, St Annes, Blackpool, Fleetwood, Barrow-in-Furness, the Isle of Man and anywhere on the Leeds & Liverpool, and a further Act of 1881 authorized the West Lancashire to purchase the branch railway. It has been suggested that the line also assisted the Leeds & Liverpool to overcome the difficulties with the tide lock; certainly the canal committee considered improving the lock and the river below it, but did nothing more. There is also some evidence that a passenger steamer may have operated on the river, when in 1882, on extending the railway to Longton, the West Lancashire opened a station called 'River Douglas' close to their swing-bridge over the river, with no access from a public road. In that year an iron paddle steamer called *Virginia* was registered by the West Lancashire at Southport, 46 tons gross, but no record of any service has come to light and River Douglas station disappeared from the timetable in 1887.[7] In 1898 the Lancashire & Yorkshire (their successors) announced that they intended to abandon the Tarleton branch on discovering that they did not own the land on which it terminated. The Leeds & Liverpool, who wanted to impose an annual tenancy at £10 p.a. in place of the peppercorn rent, agreed to sell and the railway company continued to operate the branch until 1930.[8]

In 1882 considerable alterations were made to the Liverpool terminal when the Lancashire & Yorkshire Railway rebuilt Exchange station, opened in 1886–8.[9] Liverpool corporation carried out street improvements in the Pall Mall area at the same time, filling in the original basins on either side of Old Hall Street and the arm to them behind Great Howard Street. A new warehouse was, however, opened at Bank Hall, a basin and sheds at Bootle, and new basins, warehouses and offices in Pall Mall with part of the money received for land from the Lancashire & Yorkshire, and the old terminus was closed in 1886.

A proposal to widen the canal between Liverpool and Ell Meadow lock, with other works, at £108,463 was rejected by the board in 1893. The Aire & Calder, however, in 1891 made one of their periodic complaints to the Leeds & Liverpool over the condition of the canal in Yorkshire, and the next year the board agreed to deepen the canal up to Shipley to 6 ft, the Aire & Calder agreeing to improve the Bradford correspondingly.

The use of steam propulsion grew steadily. A steam tug was ordered in 1875 and a new steam committee barge in 1878, which was so successful that G. Wilkinson & Co, who had fitted the engine, were asked to prepare plans for converting the Wigan–Liverpool coal barges. In 1880 four steam fly-boats were ordered, and a tug for Foulridge tunnel where 90 boats were being legged through each week. It seems that legging still continued, however, probably because it was cheaper and the tug's timetable not always convenient, for in 1882 a legger was suffocated in the tunnel, and £50 was granted to his widow. Thereafter legging was prohibited. Gannow tunnel was provided with a tug in 1887.

In 1891 the two odd shares in the Douglas Navigation still owned by Edward Holt were purchased for £2,000. To finance improvements and additional water supplies, £40,160 worth of mortgage debentures at 3¼ per cent were issued in 1890.

Meanwhile, on the Lancashire side, a new reservoir was opened in 1885 at Barrowford to retain surplus water from the Foulridge reservoirs. The importance of new reservoirs was emphasized by severe droughts in three successive years, 1883, 1884 and 1885, followed by a fourth, the most severe, in 1887. In each year the summit was closed (three times in 1883), extending in 1884 on each side to Gargrave and to Wigan for nearly ten weeks. Water was obtained temporarily from Blackburn corporation in the latter year, and on a long-term basis from 1885.

The stoppages drew acid letters from the Aire & Calder, who pointed out that current rates left no margin for losses due to low water, and asked what was being done. This was in March 1885, and in September they wrote again asking for a reduction in tolls, which was refused. Then in 1887 less than half the normal winter and spring rainfall closed the summit from mid-July to mid-September; well over half the canal was closed to traffic for fully a month, and much of it longer, resulting in losses of between £2,200 and £2,500 a week. From 1884 onwards a great deal of money was spent on a large new reservoir near Winterburn, in a narrow valley about 5 miles north of Gargrave which impounded the upper reaches of Eshton Beck. Edward Filliter, the Leeds consulting engineer, estimated the cost of a reservoir holding 284,000,000 gallons at Winterburn at £45,000, and the work took 9½ years to complete. Delay was occasioned by the need for an Act to secure water rights which it had been hoped could be obtained by agreement. It was passed in 1891,[10] and also authorized construction of a pipeline to discharge at the head of Greenberfield locks. The pipeline, 8⅞ miles long, capable of carrying 9,750,000 gallons per day, cost £48,975, while the reservoir amounted to £84,765, totalling £133,740. This was the last major construction work undertaken by the company, and brought the total capital expenditure on the canal to £1,684,062.

The formal opening was conducted with due ceremony on 17 August 1893 by the Chairman, Alfred Harris, accompanied by members of the board, senior officials, and Henry Rofe, who had succeeded Filliter as consultant. The party proceeded to Foulridge,

> '... and thence on the Steamer "Water Witch" to Greenberfield where Mr Fotherby, Contractor, and Mr A. W. Stansfeld (assistant engineer, eastern side) joined them. The valve of the pipeline from the Winterburn Reservoir was then opened by the Chairman and the water allowed to flow into the summit pool, the Chairman having previously made a few appropriate remarks. Mr Rofe presented the Chairman with a silver key, enclosed in a handsomely ornamented silver box, with which to perform the opening ceremony'.[11]

Even Winterburn could not prevent droughts, however, and after shortages in 1898 Rofe was asked to report on a second reservoir in the valley. He advised that one could be made above the existing one for £126,000, but apart from some trial borings nothing more was done.

The fall in dividend from 17 per cent in 1890 to 5¼ per cent in 1894 naturally caused great concern, and in February 1894 a sub-committee was appointed to enquire into the financial position. They reported that estimated expenditure on continuing improvements over the next few years would be £10,000 p.a., which would exhaust the company's resources by 1897 without extra revenue. Although receipts over the previous five years had increased, the carrying profit had fallen owing to wage increases and more traffic having to be sent by fly-boat to compete with the railways. Rates and taxes had gone up, partly due to increased warehouse space, and there had been the heavy cost of the new reservoirs. But apart from economies 'to be considered' in the Engineering Department, the sub-committee could only suggest increases in revenue as a solution, but with no concrete proposals as to how it should be obtained.[12] In terms of boats, the increase in trade on the canal since 1890 had been:

	1890	*1894*
Private traders' craft	771	894
Company's craft	98	130
" steamers (for carrying)	16	26
Total	985	1,050[13]

A prolonged frost early in 1895 worsened matters. Urgent traffic from Liverpool was sent by rail and for five weeks the canal from Wigan to Armley was closed, including the Leigh and Rufford branches, except for the length between Nelson and Riley Green where condensation water from the mills prevented thick ice and the coalowners had strengthened their boats. Arrangements were made with the Midland Railway to take some Yorkshire traffic at Colne and boatmen and maintenance gangs were put on half pay. As a result, £4,144 had to be transferred from the contingency fund to revenue. The whole of the debenture stock of £320,000 had by now been created and allotted, and the board authorized creation of a further £50,000 worth. The company ceased to work the Skipton quarries, and they were leased to a new company formed for the purpose, the Skipton Rock Co Ltd. The lease of Rain Hall quarry had been relinquished in 1891, when it was worked out.

In 1896, discussions were commenced with Bingley Urban District Council on a scheme by A. W. Stansfeld (now director) and W. N. Harris for generating electricity by water power from the canal, presumably using the fall of the five- and three-rise locks, but agreement on charges for power could not be agreed and Bingley decided to enlarge its gasometers instead. The canal company decided to create £275,000 additional authorized capital under the 1891 Act, and issued £122,500 preference stock at 7 per cent.

Improvement works slowly proceeded, reconstruction, widening and deepening on the main line and subsidence correction on the Leigh branch being authorized to the tune of £17,600 in 1899, while R. H. White, the engineer, estimated that completion of work already started on the Barrowford–Blackburn pound would cost another £44,357. As the capital raised by the preference issue had already been overspent by £32,384, the board decided to issue the balance.

To add to their troubles, the Lancashire & Yorkshire and London & North Western railways announced in 1898 that they proposed to reduce grain charges from Liverpool to Blackburn, the subject of an agreement with the Leeds & Liverpool, by 1s (5p) per ton on shipments over 50 tons. The canal company pleaded for 6d (2½p), whereat the railway companies stated that they considered the rates agreement terminated.

Early in 1900 the Ince Hall Colliery branch was closed and filled in following a burst, and the basin for Wigan power station was constructed. Two bursts caused by subsidence on Ince Moss, on the Leigh branch, resulted in the construction of Poolstock lock. Despite the numerous expedients of the preceding years, the ordinary dividend in 1901 was down to 1 per cent. A committee of shareholders was formed at Bradford in 1901 to investigate the company's affairs. In the spring of 1902 they reported. They considered the condition of the canal was a credit to the engineer, although their consultant thought £500 a mile for improvements over 58 miles from Liverpool was 'somewhat excessive' and further work should be open to tender. More might have been done to conserve water, but on the other hand the advance treatment of sections subject to subsidence was praised. Management, on the other hand, was severely castigated for letting opportunities slip for gaining new traffic. The committee said it

> 'holds strongly that a policy of "not considering things worth while" is absolutely fatal. . . . A canal capable of only dealing efficiently with early 19th century traffic is not in a position to do really well under existing conditions, but, under more energetic management, better results could no doubt be obtained'.

Private traders should be encouraged, they thought, and an able traffic manager appointed.

Shortly afterwards R. M. Stansfeld was appointed to the board and his father, A. W. Stansfeld, the vice-chairman, was made managing director. In November 1904 A. W. Stansfeld suggested that as he gave his entire time to the company's affairs he would prefer to become paid general manager. The board, although acknowledging that his activities had been beneficial (after two years with no ordinary dividend, 1 per cent was now being paid again), preferred him to retain his present position and granted him a salary increase. But when R. H. White, the engineer, died in May 1907 Stansfeld resigned his directorship and became general manager and engineer.

Between 1902 and 1904 several moves were made towards enlarging the locks from Wigan to Blackburn to the same dimensions as those below Wigan. An estimate of £24,000 was prepared, but nothing was done beyond the rebuilding of Ell Meadow lock, authorized in 1903, to counteract subsidence, followed by the building of Pagefield lock in 1904 for the same reason. Yet subsidence was so rapid that the walls had to be raised again in 1909. Despite the new reservoir, a prolonged dry season in 1901 caused the closure of the summit level, during which opportunity was taken to repair Foulridge tunnel. The repairs evidently were not very effective, as in March 1902 a section of tunnel near the east end collapsed, completely blocking the waterway and a road above, closing the canal for three weeks.

The need for revenue was acute. After nine years of low ordinary dividends (1 per cent except for 1908 when it rose to 1½ per cent and 1909 when it dropped to ¾ per cent), 1913 once more saw no dividend declared and this continued throughout the war until

1919. Water shortages and subsidence on the Leigh branch continued to cause expenditure. There was a partial stoppage from drought in 1911 when water was pumped into the Leigh branch from flashes in the neighbourhood, and at the year end the board decided to alter a number of Wigan locks in order to equalize them to save water. The work was completed in 1913, when there was another serious drought, and yet another in 1915. In 1912 an effort was made to attract a new traffic by developing Church wharf for handling textile machinery made in Accrington. Then in 1913 subsidence caused Poolstock lock to be rebuilt. Wartime inflation, too, added to the depressing financial picture. Boatmen's wages were increased, the Works Department was given a war bonus, although the company managed to increase the Foulridge tunnel dues. In 1917 the canal was placed under government control through the wartime Canal Control Committee. Compensation was paid based on net average revenue over the five years ending December 1913, and when this came to an end in 1920 the company's finances were in a parlous state. Merchandise carrying was discontinued in 1921, boats and steamers were sold or leased, warehouses were let and wages reduced. Dividends for 1919–23 were:

1919	½ per cent	1922	Nil
1920	1 per cent	1923	1 per cent
1921	Nil		

In 1921 and 1922 no preference dividend was paid.

R. M. Stansfeld died in 1924 and A. W. Stansfeld, engineer and general manager, retired. He was retained as consultant and Robert Davidson was appointed engineer and general manager in 1926.

Further extensions were made to Shipley wool warehouses in 1923, 1927 and 1928, and at Church in 1925. A publicity booklet on the canal was produced in 1926 in an endeavour to attract traffic, and in 1928 the company made a profit of £2,394 on constructing a short branch canal for the new power station at Armley, near Leeds. In 1929 the overhead power line gantries were built along the canal from Stanley Dock locks to Litherland, which became so familiar a part of the skyline in north Liverpool.

In 1930 the company made its last attempt to make a profit from carrying by taking shares in an amalgamation of four carrying concerns, Lancashire Canal Transport Ltd, Benjamin C. Walls Ltd, John Hunt & Sons (Leeds) Ltd, and Liverpool Warehousing Co Ltd. The Leeds & Liverpool provided the manager and secretary, and half the £40,000 capital of the new concern, Canal Transport Ltd, which leased Shipley warehouse and a new shed built for the purpose at Leeds, and later the Blackburn, Church, Burnley and Wigan warehouses. A conference was held with the London, Midland & Scottish Railway to avoid undue competition and indiscriminate rate-cutting, although the carriers were forced to reduce their Liverpool–Shipley wool rates by 1s (5p) a ton when the railway reduced theirs.

In 1933 the Wigan fitting shop and boatyard were let to James Mayor & Co Ltd of Tarleton, in which firm the company took shares and were represented on the board. Further shares were acquired later. Mayor & Co were still making a small profit in 1945, so the Leeds & Liverpool's investment produced a little income. The same cannot be said of the Canal Transport investment up to March 1940, when a loss of £2,392 was recorded. During the next four years carrying profits rose sharply as wartime traffic increased, but in 1945 there was a loss of over £10,000 which continued until nationalization.

In 1935 two directors advocated enlargement of the locks on the Yorkshire side to accommodate two boats at a time and speed up traffic, but the cost was too great. Revenue deficits occurred in a number of years during the thirties—in 1937, following purchase of new boats, and the same again in 1939. Ordinary dividends from 1925 to nationalization were:

1925	1 per cent	1940	$\frac{3}{8}$ per cent
1930	$\frac{1}{2}$ per cent	1945	$\frac{3}{8}$ per cent
1935	$\frac{1}{2}$ per cent	1947	$\frac{3}{8}$ per cent

Although the war brought increased traffic, the canal suffered considerably from enemy action, particularly on Merseyside, including the destruction of the company's head office.

Water shortage continued whenever there was a particularly dry summer—1937 was a bad year in this respect. Then in 1945 a bad breach occurred in the Leigh branch from subsidence and another serious stoppage from drought in 1947, when the summit and the locks at Blackburn, Johnson's Hillock and Wigan were closed for five weeks.

When the canal was nationalized on 1 January 1948, the deficit was £22,151 on revenue account, and £21,244 in respect of Canal Transport Ltd. The last general meeting was held at the Midland Hotel, Manchester, on 24 March 1948, and after declaring the final dividends the balance of £25 was presented to the chairman,

R. W. Wickham, JP, 'for the purchase of some momento of the old days and as a small tangible token of appreciation for services rendered'.[14] R. Davidson, the general manager, became a member of the Docks & Inland Waterways Executive.

Under the Executive the canal continued to be maintained in good condition except for the little used Rufford branch, which deteriorated somewhat. But traffic continued to decline, and in 1963 British Waterways, who had acquired Canal Transport Ltd's fleet, ceased most carrying as part of a national policy. On the Leeds & Liverpool commercial traffic had, in any case, all but ceased. The last traffic over the summit, occasional sugar boats from Leeds to Liverpool, passed in 1960, when the last known horse-drawn boat also disappeared. The last steamers ran in the mid-fifties. The final commercial traffic on the main line was coal, between Burnley and Whitebirk power station, Blackburn, in the spring of 1963, and between Wigan and Atholl Street gasworks in January 1964. Coal traffic on the Leigh branch to Wigan power station still continues at the time of writing.

Today the canal is in fair condition and, with the exception of 33 chains at the Liverpool end, closed in 1960, is still navigable throughout. The terminal section has been filled in and the canal terminates at Chisenhale Street bridge.

Pleasure traffic growth has been slow, due to the long distance and heavy lockage in reaching the scenically very fine section between Apperley Bridge and Burnley from the midland canals, but several hire firms are now established on both sides of the summit.

The Springs Branch or Lord Thanet's Canal

In 1846 two improvements were made to the tramroad. The long drop from the staithes having caused continual damage to boats, the old ones were superseded by a 200 yd self-acting rope-worked incline down to the canal head, where short chutes were installed. At the quarry end of the line, a 250 yd diversionary tunnel was cut to even out the gradient and from 1846 a steam winding engine was used to replace horse haulage. So there were now two inclines: the short, steep, self-acting one from the canal staithes and the longer, more gradual ascent worked by the stationary engine. Locomotive haulage replaced the latter in 1892.

In 1896 the then partners in the quarry sub-lease formed themselves into the Skipton Rock Company, and the Leeds & Liverpool agreed to the conversion of the tramroad to a railway of main-line dimensions, so that it could be connected to the Midland Railway's sidings at Embsay on the Skipton–Ilkley line which had been opened in 1888. At the same time a diversion was made to avoid the tunnel. Stone ceased to be shipped by canal about 1947 and the incline rails were removed, although the quarry lines continued to be connected to the main-line railway at Embsay. In an attempt to revive canal carriage, the track bed was levelled in 1951 to allow lorries to bring stone down to the chutes, but this was found to be uneconomic and was discontinued after a few weeks. From August to December 1964 the tramroad site was again used by lorries to bring down stone for use by British Waterways in making concrete piles. Since then the branch has been disused except for exploration by the occasional pleasure craft and as a feeder from Eller Beck.

The Bradford Canal

With the rapid growth of Bradford, the canal basin at Broadstones was soon close to the centre of the town instead of on its edge, while the canal's water supply became more and more polluted from domestic sewage and use and re-use by mills and dyehouses. To safeguard their supply the canal company had spent considerable sums between 1775 and 1798 in purchasing land to gain the water rights. Further, at some time the company dammed Bradford Beck and culverted its water into the canal, although the 1771 Act authorized the taking of water only from its tributary, Bowling Mill Beck. In 1844, the Board of Surveyors commented:

> 'In summer time the water is low, and all this filth accumulates for weeks or months above the flood-gates, and emits a most offensive smell. This noxious compound is conveyed through the sluice into the canal.'

There it was abstracted by canalside mills for boiler feed and cooling water, being returned to the canal, the waters of which

> 'are scarcely ever cool in summer, and constantly emit the most offensive gases. The public health suffers considerably in consequence of the neighbourhood of the Canal'.[15]

Four hundred and six deaths in Bradford during the 1849 cholera epidemic brought matters to a head. In the face of public outcry the city council agreed to prepare an Improvement Bill to deal with the appalling sanitary conditions in the borough, including the purchase of the canal for £100,000 along with the gas and waterworks, to be raised by public loan; powers were sought to close the canal after purchase and sell the land. But the canal's supporters were too powerful and the opposition of the Leeds & Liverpool, Aire & Calder, and a majority of shareholders, backed by quarry-owners and ironmasters, to the 1850 Bill forced the corporation to drop the clauses for purchase and sale and reduce the loan to £50,000. Thus, by not opposing the powers for closure but concentrating on the means, the proposal was cleverly defeated, taking care to leave unchanged the powers enforcing cleansing of the beck. The canal committee, it seems, favoured selling out to one of the railways then thrusting into the West Riding, but were overruled.[16] The Leeds & Bradford Railway, opened in 1844, competed directly with the canal by running parallel to it.

The canal company had leased the canal for varying periods from 1788 onwards, usually to groups of stone and lime traders, and from 1852 to Jeremiah Crowther and Samuel Dixon, limeburners.[17] Apparently the previous lessees had found it unprofitable, as shortly before expiry the company, probably seeking better terms, offered it to the Leeds & Liverpool for 7 years at £1,400 p.a. The Leeds & Liverpool consulted the Aire & Calder, who declined to take a joint lease, but the two companies then agreed with Crowther & Dixon to allow £450 p.a. from the Leeds & Liverpool on limestone and £200 p.a. from the Aire & Calder on merchandise, flags and slate traffic provided the lessees did not sub-let without their consent, did not raise their tolls, and kept the locks sufficiently full to take Leeds & Liverpool boats.[18]

Meanwhile, agitation against the canal nuisance continued to grow, and the hot summer of 1864 finally provoked a group of private individuals to open a fund to seek a court order against the company compelling it to abate the public nuisance from what *The Bradford Observer* called 'that seething cauldron of all impurity, the Bradford Canal'. The company pleaded that they were powerless to prevent the nuisance as the beck water was already heavily polluted before it reached the canal, but the judge decided in favour of the plaintiffs, who promptly obtained an injunction in March 1866, restraining the company from diverting Bradford Beck into the canal on penalty of £10,000, from 6 November that year.[19] The Leeds & Liverpool in the meantime offered to take over Crowther & Dixon's lease for a year in an effort to purify the canal, rather than lose the traffic, but as the plaintiffs would not agree the Bradford company was forced to refuse.[20] The Leeds & Liverpool, of course, was then itself leased to the railways whose increased tolls had contributed to the Bradford's falling merchandise revenue.

The Bradford company now decided to close the last ¼ mile of their canal above Northbrook bridge and sell the land, and tried to persuade the Leeds & Liverpool to lease the remainder. The latter preferred to take the entire canal as a going concern and, if land had to be sold, reap the benefits themselves. They saw no attraction in acquiring a truncated canal needing not only new wharves and warehouses, but back-pumps for supplying water from their own canal, too. Quite firmly they stated that they would neither lease nor purchase without the head-level intact. In November two Bills were deposited, one by each company, setting out their respective positions, and in order that they could be heard the Bradford company obtained a stay of the injunction until May 1867. The preamble to neither Bill being proved, both were rejected, and on 1 May 1867 the Bradford Canal was closed and drained.[21]

The quarry owners and stone merchants of the district now found themselves without suitable transport, as in most cases it was not practicable to use the railway, and they opened negotiations with the Leeds & Liverpool, Aire & Calder, and Bradford corporation. The Bradford Canal Company, ready to make a deal with anyone, agreed to sell the main part below Northbrook bridge to a new company for £2,500, the portion above to be sold to builders for an unspecified figure. The corporation received land for widening Canal Road and £2,000 from the purchasers towards culverting the beck, and the old company was wound up on 21 March 1870 under a Chancery Order which directed that an Act should be obtained. This was done in 1871[22] and a new company, The Bradford Canal Company Limited, registered with a capital of £35,000. Nine stone merchants, four coal and lime merchants and a boat builder, all from Bradford, Shipley and Leeds, were directors. The prospectus noted that the average tonnage during the three years up to closure had been 125,000 tons p.a., and confidently predicted that it could quickly be increased to 180,000 tons. Water could be obtained from three streams and two reservoirs for nine months of the year, and by back-pumping from the lower pounds during the other three at an estimated cost of £650 p.a.[23]

An abortive request was made to the Leeds & Liverpool to supply water by gravity through a pipeline from the head of Bingley locks,[24] after which steam pumping engines were erected at each lock to back-pump from the pound below, the lowest one drawing from the Leeds & Liverpool. On the fifth anniversary of the closure, 1 May 1872, the canal was reopened as far as Oliver lock, and on 15 April 1873 a grand ceremonial reopening of the whole distance took place with a hired steam tug hauling a stone-barge full of guests.[25] The canal's shortened length was 3 miles from Shipley.

But five years of closure had had the inevitable effect. Although an outlet had been secured for the stone traffic, other regular users had turned to the railways and mineral traffic alone was insufficient; £25,000 was spent on restoring the canal, erecting wharves and accommodation, and in 1874 a mortgage of £5,000 was raised by the new owners.[26] In 1874 they approached the Leeds & Liverpool, who first agreed to lease the canal, but then to purchase it in equal shares with the Aire & Calder for £27,000 and pay off the mortgage. The Bradford Canal (Transfer) Act of 1878[27] authorized the sale and the second Bradford company was wound up. The Act authorized a joint managing committee of three members from each of the new owners, the chairman to alternate yearly, and the Leeds & Liverpool agreed to supply water for pumping free of charge, and to manage the canal.

The venture was too late to be a success. Three large warehouses were built at the Bradford terminus, which afforded the best warehousing facilities in the town for a time until they were rivalled by the equally large and up-to-date premises of the Lancashire & Yorkshire and Midland railways. Capital expenditure by the joint committee, including the purchase price, amounted to £58,437 in 1892.[28]

Tonnages rose, and the terminal accommodation had to be enlarged at a cost of £12,000 in 1895,[29] and again in 1902:[30]

Year	*Tons*
1888	80,674
1898	96,037
1905	98,271
1910	102,390

About two-thirds of the traffic came from the Aire & Calder and one-third from the Leeds & Liverpool, much of the latter being wool and sugar from Liverpool. As late as 1914 a fly-boat was still sailing to and from Liverpool once a week. But much of the canal's income came from warehouse rents on goods not carried by water, with a gradually widening gap. Consequently, income did not rise fast enough to keep pace with costs, particularly pumping expenses; therefore profits were only marginal in the years when losses were not actually being made. In 1897, when the steam pumps were nearly life expired, replacement by electric pumps was considered, but the cost was felt to be too great.[31]

In 1902 the Aire & Calder proposed enlargement of the Bradford to take their larger boats and so avoid transhipment at Leeds, but the Leeds & Liverpool were unwilling to widen the canal between Leeds and Shipley. Tonnages began to fall after 1910 and during World War I deficits were continuous, less than 150 boats using the canal from 1917 to 1922. The cost of pumping was estimated at £8 10s (£8·50) per boat, and in 1921 the Aire & Calder were finding it cheaper to carry by road from Leeds to Bradford.[32] In 1920 tonnage carried was down to 38,821.[33] The financial picture from 1888 to 1922 was as follows, excluding interest on the joint companies' investment:

Year	*Tolls*	*Total Income*	*Pumping Costs*	*Total Expenditure*	*Surplus (+) Deficit (—)*
	£	£	£	£	£
1888[34]	1,143	2,484	1,165	2,182	+ 302
1892	1,080	2,379	1,406	3,003	— 624
1898[35]	1,030	3,019	1,544	2,747	+ 272
1903[36]	1,076	2,967	1,257	2,634	+ 333
1908	1,020	2,984	1,486	2,548	+ 436
1916	579	2,828	1,765	3,053	— 225
1921	356	2,867	3,467	5,774	— 2,807
1922	325	2,465	1,039	4,633	— 2,168

Total capital invested by 1921 amounted to £82,500.[37]

In 1920 the Leeds & Liverpool suggested to the Aire & Calder that the canal should be closed, and in 1921 an abandonment Bill was deposited. Bradford corporation, the Chamber of Commerce and West Riding County Council opposed as a matter of public policy and the preamble was not proved. A second Bill in 1922 was successful and the Bradford Canal (Abandonment) Act became law in June that year,[38] authorizing the joint owners to close the canal and dispose of the land, sharing the proceeds equally. The canal finally closed on 25 June 1922. Only the 2 furlongs from the junction to the foot of Windhill lock remained for use as moorings, the remainder being sold, filled in or just left. In close on fifty years since closure the remains have progressively become less, and today the last remnants are disappearing fast.

CHAPTER XVII

The Lancaster

HAVING taken on the lease of the Lancaster & Preston Junction Railway[1] on 1 September 1842 for two years at £13,300 p.a. (representing 4 per cent dividend), plus payment of the interest on the railway company's debts of £113,000, the Lancaster Canal company undertook not to oppose any northward extension of the railway. The Lancaster & Preston Junction obtained an Act of authorization in 1843.[2] In that year two schemes for lines between Lancaster & Carlisle were mooted, one via Kendal and one via the Lune valley. The canal company supported the Lune valley route, which would not directly compete, to the extent of promising to invest £50,000. However, in order to buy off opposition to the railway company's leasing Bill they were forced to withdraw their offer.

On acquiring the railway's locomotives the canal company agreed with the Manchester, Bolton & Bury Canal Navigation & Railway Co to work the latter's traffic to Preston in conjunction with their own, to compete with the North Union Railway for the Preston–Manchester traffic. The Carlisle railway scheme had crystallized in 1843 into a route via Kendal and early in 1844 the canal company made a provisional agreement with the Lancaster & Carlisle Railway whereby the canal would, from 1 January 1846, maintain and manage the North End and Glasson branch 'for and on behalf of' the railway company in return for £11,000 p.a. guaranteed income. The second half-year's receipts for 1843 had been £10,223, so the company evidently anticipated a serious fall when the railway opened. Simultaneously the Lancaster & Carlisle was offered an option on the unexpired term of the Lancaster & Preston Junction lease for £1,250 p.a., stone for construction from the canal company's Lancaster quarry, and general assistance in building the line. But disagreement over the inclusion of Preston wharves in the agreement caused the canal company to oppose the

Lancaster & Carlisle's Bill. They were unsuccessful, and the railway was authorized in June 1844.[3]

While these affairs were going on, the canal company was operating the railway, very successfully in their own view, but hardly so in the eyes of the public. Their first act had been to discontinue the Preston–Lancaster packet boats and simultaneously raise the railway fares, removing the seats from the third-class carriages in order to accommodate more passengers standing than sitting. Then in 1845 they took over goods carrying from Hargreaves & Son, who had acted as carriers on the Lancaster & Preston Junction as on other Lancashire lines, thus creating for themselves a complete monopoly of goods and passenger traffic by rail and mineral traffic by canal which they exploited to the full, making in that year £5,000 profit from the railway, which increased considerably in each succeeding year.

At the same time relations with the lessors deteriorated to rock bottom. The canal company had been dilatory in preparing a lease, so that when they presented the draft to the Lancaster & Preston Junction in October 1843 there had been a partial change of heart following the promotion of the Lancaster & Carlisle. Consequently the former's board took the opportunity to object to clauses permitting a sub-lease, which the canal company probably had inserted for the same reason. The leasing Act unfortunately gave only general powers without setting out exact terms. Then in October 1844 the Lancaster & Preston Junction's board decided to lease the railway to the Lancaster & Carlisle instead, to which the shareholders agreed, considering themselves free to do so as agreement with the canal company was still not concluded. But while details were being thrashed out between the two railway boards, the 'little' North Western Railway (so called to avoid confusion with the London & North Western) was incorporated in 1846 to build a line from Skipton to Lancaster and Morecambe. The canal company had co-operated to the extent of allowing it to pass beneath their Lune aqueduct, but the Lancaster & Preston Junction board considered it a threat to their line, and were precipitated into outright amalgamation with the Lancaster & Carlisle. But when they put this to their shareholders for formal ratification in February 1846, the proposal was unanimously rejected on the grounds that better terms could be obtained by waiting. All but one of the directors thereupon resigned, and as the meeting had no power under their Act to appoint a new board the company found itself in the astonishing situation of having no directors nor means of appointing any, and a railway which in theory was legally leased to the Lancaster & Carlisle Railway, but in practice illegally leased to the Lancaster Canal, who, regarding possession as nine points of the law, continued to run the trains. To add a further complication, the canal company was still trying to persuade the Lancaster & Carlisle to take over its traffic under the 1844 agreement.

Demand was followed by counter-demand. In November 1845 a formal notice served on the Lancaster & Carlisle by the canal company, indicating that they would consider the 1844 agreement in force from 1 January 1846, was returned with the demand that the Lancaster & Preston Junction be handed over to them in accordance with their 1844 lease. The canal company replied by starting Chancery proceedings to obtain an injunction against the Lancaster & Preston Junction for enforcement of the 1842 lease. The confusion was heightened in September 1846 when the Lancaster & Carlisle not only opened their line from Lancaster as far as Kendal, but calmly ran their trains through to Preston over the Lancaster & Preston Junction (to which they had constructed a link at Lancaster despite canal opposition), without authority and ignoring the canal company's protest. The canal immediately withdrew their Lancaster–Kendal packets, but continued to run local trains on the railway. The Lancaster & Carlisle rendered statements of their traffic over the Lancaster & Preston Junction, but refused to pay tolls on the grounds that there was no legally constituted body to which to pay them. Then in February 1847 the East Lancashire Railway stepped in and offered to buy the canal and the Lancaster & Preston Junction. Terms were agreed at £23,500 p.a. for the canal, redeemable within ten years at 25 years' purchase, and £29,000 outright for the railway. Again agreement was very close when both parties withdrew after the railway's shareholders appointed a new (strictly illegal) board to consider the proposal. The canal company tried again to obtain performance of the 1842 lease, in reply to which the railway shareholders unanimously rescinded their original resolution approving it.

The canal and Lancaster & Preston Junction committees then met, the latter determined not to lease to the canal, which, in turn, was prepared to relinquish the railway only on obtaining payment of adequate compensation, execution of their 1844 lease to the Lancaster & Carlisle if possible, and support from the Preston Junction in obtaining payment of tolls from the Carlisle. The Lancaster & Preston Junction refused, so the canal company withheld the half-year's rent due in March 1848 and turned to the

Lancaster & Carlisle with whom, surprisingly, relations were still relatively friendly despite the argument on tolls. In place of the 1844 agreement they offered to sell their canal to the railway company outright for 38s (£1·90) per share and 5 per cent p.a. in perpetuity, complete with the Lancaster & Preston Junction, on payment of the outstanding tolls, the amount of which was still subject to negotiation. The Carlisle company counter-offered 30s (£1·50) subject to the railway company taking over the application for an injunction against the Preston Junction and successfully obtaining it. The canal company accepted, and then yet again the negotiations fell through when the Lancaster & Carlisle wanted to make unacceptable changes.

Both companies continued to run their own trains over the Preston Junction until, inevitably, farce ended in tragedy. On 21 August 1848, a Euston–Glasgow express hauled by a Lancaster & Carlisle engine ran into the back of a Lancaster & Preston Junction local train standing in Bay Horse Station, fatally injuring a passenger. Captain Laffan, the Board of Trade inspecting officer, severely criticized 'the want of proper understanding between the Lancaster & Carlisle and the lessees of the Lancaster & Preston Railway' at the ensuing enquiry, and urged that an agreement should speedily be reached. The Railway Commissioners told the Preston Junction to put their house in order and in December a special meeting of the canal proprietors ratified a final agreement with the railway company. The latter undertook to pay the canal £4,875 p.a. for the unexpired portion of the lease, with an option to consolidate the full sum plus 5 per cent within eighteen months, or 4½ per cent at any other time during the term. The canal company agreed to withdraw its injunction proceedings, each side bearing its own costs, and the Lancaster & Preston Junction promoted a Bill to gain parliamentary sanction. If the Bill were to fail the railway agreed to ratify the 1842 lease to the canal.

The Bill did not fail. In fact, the Act of 1849[4] went a stage further and vested the Preston Junction in the Lancaster & Carlisle on a profit-sharing basis following a further, independent, agreement between the two railway companies. The Carlisle company were ordered to pay outstanding tolls to the canal company on terms to be settled by arbitration, at which Robert Stephenson awarded the canal £55,552. The canal gave up possession of the Preston line on 1 August 1849.

Despite the incredible seven years' legal wrangling and the low ebb to which it had let the railway service fall, the Lancaster Canal company seemed to escape public odium, the full force of which had been directed at the unfortunate Preston Junction. Indeed, the canal company came out of the affair with a profit of £67,391, which enabled them to pay off their outstanding mortgages of £26,000, give each proprietor a bonus of £1 17s 6d (£1·87½) per share and establish a £6,700 contingencies fund; the dividend had moved from 2 per cent in 1845 to 2½ per cent in 1846, 2¼ per cent in 1847 and 1848, and then back to 2 per cent in 1848 and 1850. Despite their consistently low dividends by comparison with other canals, the company remained a considerable power in north Lancashire, as was shown by the delicacy with which the Lancaster & Carlisle made their approaches, and the strength they still possessed in 1850, when a traffic sharing agreement was concluded between the two companies. The Lancaster & Carlisle took the passenger and merchandise traffic to Kendal, and the canal retained the coal and heavy goods; it also continued to carry between Glasson Dock and Preston. Coal rates for destinations beyond Kendal were to be subject to agreement. The establishment had been bent but it was still far from broken.

Before leaving railway and canal politics, it is necessary to return to events leading up to the authorization of the 'little' North Western Railway in 1846. Although Glasson Dock was busy, ships entering it still had to contend with the sandbanks of the Lune estuary. As ships grew in size, the difficulties increased, and all the while there was a threat from the spasmodic improvement of the Ribble. A new navigation company had been incorporated in 1838 to cut a deeper channel, enlarge Preston quays, and construct a dock at Lytham where large ships could safely discharge into lighters; this was completed in 1842.[5] Simultaneously, competition from a new source was presented by Fleetwood, to which the Preston & Wyre Railway was opened in 1840 to establish a port there. Consequently, three schemes were being canvassed in Lancaster: deepening of the channel to the town's quays; construction of a new dock at Thornbush (site of the 1799 scheme) and a railway to the Lancaster & Preston Junction at Ellel; and docks at Poulton Ring (now Morecambe) with a 3½ miles long ship canal across the peninsula to the Lune where a dam across the river would maintain deep water at Lancaster quays.[6] The last scheme was probably one reason why Lancaster was chosen as the original terminus of the 'little' North Western Railway, of which Edmund Sharpe, a Lancaster engineer and architect, was secretary, the intention being to use an improved port of Lancaster as a means of supplying the West

Riding of Yorkshire from the west coast. In securing Admiralty consent to bridging the Lune, the Lancaster & Carlisle Railway paid £10,000 to be used for the improvement of the river, and £16,000 compensation to Lancaster merchants, so there was now capital available to finance a scheme. In 1845 the Tidal Harbours Commission held an enquiry into the best method of laying out the £10,000, at which Sharpe attacked the Thornbush scheme as a further blow to the port of Lancaster by the canal company who, as lessees of the Preston Junction line, already monopolized the town's communications, and went on to press for the ship canal scheme on behalf of the 'little' North Western (backed by the Midland Railway with which it would connect at Skipton), Poulton landowners and Lancaster merchants. The Lancaster Canal and the Lancaster & Carlisle Railway opposed it, on the grounds that the future of Glasson Dock would be severely prejudiced (the Lancaster & Carlisle, of course, being interested in the canal under the unratified 1844 agreement).

Shortly afterwards The Morecambe Bay Harbour Company was floated to make a harbour and ship canal to Lancaster, with a capital of £300,000, but a few days later, in November, the formal prospectus was found to substitute a railway for the canal. According to Sharpe, the opposition of the Lancaster & Carlisle had not been anticipated, and as they were giving £10,000 for improvement of the Lune it was likely that their opposition would carry weight in Parliament. Furthermore, the estimate for a canal scheme was £50,000 more than had originally been supposed, which would make shipping charges unduly high. There was some short-lived agitation in Lancaster for a canal, and some of the merchants' support was lost, but in 1846 the Morecambe Harbour Railway Act was passed with little opposition, and later in the year the 'little' North Western exercised their option to purchase the company. The Lancaster & Carlisle duly paid over their £10,000 and the Lune was deepened, but the improvement was temporary and Lancaster gradually all but succumbed as a port to Morecambe which, under the Midland Railway, finally transferred its shipping activities to Heysham in 1904.

In recognition of his negotiations with the railway companies in 1844, B. P. Gregson's salary of £1,000 p.a. was fixed as a minimum for the remainder of his service. This did not stop him from accepting the management of the Edinburgh & Glasgow Railway in 1846. His new employers were agreeable to his continuing to work for the canal company, who were glad to accept his dual role at an undiminished salary. His father died in October, aged 83, after 54 years' service and B. P. Gregson was appointed clerk in his place. Samuel had been to the Lancaster what Priestley was to the Leeds & Liverpool, one who assumed duties far beyond those for which he was paid in an endeavour to promote the prosperity of his company. A highly respected figure in Lancaster, he was twice mayor.[7]

The 21 years' lease of the merchandise tolls of the South End to the Leeds & Liverpool for £4,335 p.a., consequent upon the railways' lease of the Leeds & Liverpool's own merchandise traffic, referred to in Chapter XVI, was ratified by a special general meeting in October 1851. Abortive negotiations took place at this time with the Kendal and Windermere Railway about a tramroad from the canal head to the proposed Kendal station, and in 1851 there was a vague proposal to expand the canal's traffic by a tramroad from Kendal to 'Staveley, Birthwaite (now Windermere) and other parts of the Lake District',[8] but nothing transpired and coal continued to be carted from the canal head to Kendal station. But a determined effort was made to counter the effect of the Lancaster & Carlisle Railway, which was eating into the Glasson trade. Long-term contracts for the carriage of coal at reduced rates were effected, and in February 1851 it was resolved to engage in the coastal trade. The schooner *Woodbine* was purchased, followed by the *Richard* in 1852 which was lost in the Duddon within six months. The *Oriental* was bought, then in 1853 the *Bloomer*; three more were ordered in 1855, and a screw steamer *Dandy* was tried in the canal in the same year. The company's coal trade with Ireland had grown to the extent of renting a quay at Belfast and appointing an agent, but the fillip to the Glasson revenue was temporary and did not arrest the gradual decline which had started in 1850.

In 1853 the Lancaster & Carlisle Railway were informed that the canal company were carrying pig iron and treacle from Glasgow through Glasson to Preston, chiefly for Blackburn and east Lancashire, at very low rates to compete with sea carriage to Fleetwood and Preston. The canal company felt that these cargoes might possibly be construed as contraventions of the agreement and laid their books open to the inspection of the railway company who, however, made no objection even when the jointly appointed referee reported that technically there might be an infringement. Again the canal books were inspected in 1856, without dispute, but in September 1858 the railway gave notice of its intention to terminate the competition agreement owing to violation in respect of the pig iron and treacle traffic. In May that year they had leased the Kendal & Windermere Railway and immediately began to impede the supply of coal to the Lake District from the canal, encouraging it to go by rail to Windermere instead, in disregard of the agreement. The canal company opened new negotiations which were abruptly broken off by the Lancaster & Carlisle, who refused to enter a new agreement while the Glasgow canal traffic continued. It seems likely that the heavy hand of the London & North Western Railway lay on the Lancaster & Carlisle in this affair. As a result of amalgamations the London & North Western by this time controlled the whole of the West Coast route south of Preston; they had helped to finance the Lancaster & Carlisle and appointed seven of their directors. They were a railway not generally given to friendly co-operation with rivals, particularly canals, and a year later they leased the Lancaster & Carlisle outright.

Accordingly, the canal company took steps to acquire traffic elsewhere and in July 1859 the Lancaster Steam Navigation company's steamer *Duchess* began twice-weekly sailings with merchandise between Glasson and Liverpool, at a hire charge of £37 10s (£37·50) per trip. Experiments were made with steam tugs on the canal and in June 1860 two second-hand ones were purchased to accompany one specially built and already at work. But the coasting vessels were being successively lost at sea until in 1861 only *Woodbine* was left, and even for her marine underwriters refused to quote terms. Trade depression did nothing to help, either, and in 1862 the dividend, which had remained constant at 1¾ per cent since 1851, dropped to £1 13s 6d (£1·67½) per cent, and to £1 12s (£1·60) per cent in 1863.

The first move to sell out to the London & North Western seems to have been made in September 1860.[9] After over two years' negotiations terms were agreed and a Bill promoted in 1863 which successfully passed through the Commons but was opposed in the Lords by the Lancashire & Yorkshire Railway, who had a part interest in the tramroad as joint lessees with the London & North Western of the North Union and wanted to participate in the canal lease. The London & North Western objected, withdrew the Bill and reintroduced it in the next session when it was successful and received Royal Assent on 29 July 1864.[10] The Lancaster Canal Transfer Act authorized the leasing of the North End of the canal to the London & North Western in perpetuity for £12,665 17s 6d (£12,665·87½) p.a. and the South End to the Leeds & Liverpool Canal company for £7,075, from 1 July, with increased maximum tolls on both. Land was allocated at Preston for the enlargement of the joint station, and the London & North Western with the consent of the Lancashire & Yorkshire Railway were authorized to close the tramroad from Preston as far as Bamber Bridge, and dispose of the site. This section had for several years been practically disused, particularly after the Leeds & Liverpool had in 1853 leased its tolls on merchandise to a group of railways (see p. 399), but there was still sufficient traffic from Walton Summit to Bamber Bridge—mainly coal for mills—to justify retention of this portion until it, too, was closed under a London & North Western Act of 1879,[11] by which time traffic had entirely ceased.

At the first general meeting to follow, in February 1865, the proprietors had to decide what to do about Gregson. He presented a long statement setting out the history of the canal during his lengthy service, concluding rather pathetically with his thanks for the committee's support during that time. After a proposal to pay him off with £4,500 was hastily withdrawn by the chairman, it was resolved that he should be kept on at his salary of £1,000 p.a., which continued until his death on 3 December 1872 after sixty years with the company. The rental from the lease allowed a regular dividend of 1¾ per cent once more, and the balance in hand of £18,239 was invested. The final act came in 1885 when the London & North Western offered to buy the canal outright in return for 4 per cent debenture stock equal to the lease rental from both ends of the canal. This represented £43 15s (£43·75) for each canal share at the prevailing prices. The proprietors agreed, and the canal was vested in the railway company on 1 July 1885 under an Act of 16 July of that year.[12] The lease of the South End to the Leeds & Liverpool was not prejudiced; they continued to pay rent to the railway. After payment of the final 1¾ per cent dividend the Lancaster company were formally dissolved on 1 January 1886, and at the last general meeting each proprietor received a bonus of 10s 9d (54p). The balance of £101 4s 10d (£101·24) was handed to the chairman, who caused a number of commemorative silver medallions to be struck.

Thanks to the foresight of the committee in selling land to Kendal Gas Company for a works in 1824, regular coal traffic from Preston was assured until the motor age, there being no rail access to the gasworks. Consequently the railway kept up a good standard of maintenance and other traffic continued, particularly coal to canalside mills in Lancaster, manure, minerals, grain, timber and chemicals from Glasson for Wakefields' gunpowder mills at Gatebeck. These largely replaced the much older mills at Sedgwick

32. Obverse of the silver medallion struck at the dissolution of the Lancaster Canal Company in 1886

and Bassingill in 1850, and were connected to the canal by a horse tramway at Crooklands wharf in 1874 where sidings were laid out. The tramway continued to the railway at Milnthorpe, however, and the canal probably carried a fairly small proportion of the total traffic in later years.[13] In 1894, Glasson still provided the cheapest route for heavy goods to Kendal, the cement for the Thirlmere waterworks going by that route although the grain trade was rapidly dwindling.[14] In 1883 the London & North Western opened a branch line to the dock from Lancaster. Management of the canal was exercised by a superintendent based on Castle Station, Lancaster. Examples of tonnages and revenue after railway purchase are as follows:[15]

Year	*Tonnage*	*Tolls*	*Net Revenue**
	tons	£	£
1888	173,882	9,181	17,289
1898	165,005	8,013	18,728
1905	130,396	6,171	13,984

* After addition of other revenue (water rents, wharfage, etc) and deduction of expenditure.

Horse haulage was always used on the canal except for a steam ice-breaker and dredger, and two Leeds & Liverpool steam tugs which were tried for two or three years after World War I. The shallow depth precluded full loading of barges, however, and in 1921 they went back to the Leeds & Liverpool.[16] They came and went by sea between Tarleton and Glasson, as indeed did all new craft required on the canal in its later years.

In 1935 a London, Midland & Scottish Railway Act[17] gave the railway powers, as successors to the London & North Western, to raise tolls and in 1939 a further Act[18] authorized closure of ½ mile at Kendal owing to leakages. About 1941–2 the canal was closed from the gasworks northward. In 1944 the railway company promoted a Bill to close a number of their canals, including the Lancaster, but opposition from local authorities and several Lancaster firms succeeded in deleting the Lancaster Canal references during the Lords Committee stage. Kendal gasworks was receiving between 6,500 and 7,500 tons of coal a year,[19] but in September 1944 this trade was transferred to road and all commercial traffic north of Lancaster ceased. The last from Preston to Lancaster, again coal, was carried in 1947.[20]

The Transport Act, 1955, authorized the British Transport Commission, which had acquired the canal on nationalization, to close it to navigation, and some 5¾ miles from the feeder at Stainton Crossing bridge to Kendal were drained owing to leakage through limestone fissures in the bed—a recurrent problem since Crosley's day—and the last 2 miles were filled in; also upwards of ¾ mile was progressively drained and partly filled in at the Preston end. A growing number of pleasure craft were coming on to the canal but rarely ventured up Tewitfield locks, beyond which a further 100 yd were drained at Burton and a pipe inserted, again due to leakage. Consequently, when the Ministry of Transport's M6 motorway extension was projected to cross the canal north of the locks with five culverts, strenuous local attempts failed to have low bridges of pleasure craft dimensions substituted. So today some 42 miles of level pound south of Tewitfield, and the Glasson branch, are retained for cruising, although the canal also performs a useful function in providing water to a chemical works near Fleetwood through a pipeline from the canal near Garstang, for which reason the section above Tewitfield was retained as a water channel.

Ulverston Canal

The new breakwater came too late to save the canal. After a peak year in 1846, tonnages dropped to nearly a third in 1848 as the Furness Railway's expansion of Barrow and Piel took effect; by May 1849, nearly three times as many ships used Barrow as Ulverston and thereafter decline was steady, accentuated by the opening of the Furness Railway's Ulverston extension in 1852 and completion of the link to the west coast main line at Carnforth by the Ulverstone & Lancaster Railway in 1857. The rise and fall of Ulverston as a port is illustrated by the following table; although the figures probably include landings at neighbouring places within the port, they indicate the trend:[21]

Year	*No of Vessels*	*Tonnage* tons	*Receipts* £	*Working Expenses* £
1830	448	29,130	500*	140*
1834	464	30,100	529	342
1840	560	36,400		
1845	912	59,280	1,252	292
1846	944 (peak)	61,360		
1849	388	25,220	547	220

* Approximate figures. Expenses exclude capital works and major repairs.

The opening of the canal encouraged growth of the town, leading to building in the Hart Street, Sunderland Terrace and Quay Street areas.[22] The population of 2,932 in 1801 grew to 5,352 in 1841.[23]

The Ulverstone & Lancaster Railway built a new basin near their bridge in 1857, which thereafter was used by most of the canal traffic.[24] Under their 1862 Act[25] the Furness Railway acquired the canal for £22,000 and the company was wound up. At this time four or five vessels a day were using the canal, mainly bringing in timber and charcoal and taking out iron ore to Saltney for the Flintshire ironworks. By 1874 the number was down to about one every three days.[26] Figures for the canal in its last years were:[27]

Year	*Tonnage* tons	*Revenue* £	*Expenditure* £
1888	3,052	87	1,287
1898	4,623	104	108
1905	3,472	79	90

In 1913 tonnages had recovered to 5,755 and receipts to £444 against expenditure of £485. Traffic fell away during the war, however, and the last craft used the canal in 1916. It was abandoned in 1945 under the London, Midland & Scottish Railway Act 1944[28] except for use by pleasure boats at the company's discretion. Today, although deserted, it is still much as it was except that the lock gates and the opening bridge carrying a branch railway to a chemical works have been fixed.

CHAPTER XVIII

The Rochdale

In 1847 the Manchester & Leeds Railway had, with the canal company's agreement, sought, but failed to get, an Act to buy it. The railway board then suggested in August that, as one canal company had power to lease another for up to 21 years, they as owners of the Manchester, Bolton & Bury Canal should in that capacity lease the Rochdale. Instead, the canal company offered a lease to the powerful Aire & Calder, who were at the time also negotiating for one with the Calder & Hebble, but they refused it, partly because their engineer had reported it to be 'in a bad state of Repair',[1] partly because they could not agree to the Calder & Hebble's terms. Rejected by the Aire & Calder, the Rochdale started to co-operate with the railway (now the Lancashire & Yorkshire) by raising their tolls on transfer traffic with the Calder & Hebble to that waterway's level from the beginning of 1848. The Calder & Hebble then accepted a railway lease.

In September 1849 the Bridgewater pressed the Rochdale to reduce tolls on iron so that the trustees could carry and deliver to places east of Castlefield, and the long-distance iron toll was then temporarily cut to ½d. A sharp contest followed between the trustees and the Lancashire & Yorkshire Railway over traffic between Liverpool, Manchester and Rochdale. At this point the energetic James Meadows was appointed manager of the Rochdale. A letter having come in from the Calder & Hebble (now released from railway control after action by the Railway Commissioners), Meadows arranged a meeting with them and the Bridgewater to discuss rate reductions that had been made by the Lancashire & Yorkshire and the London & North Western. The result was agreement that Loch of the Bridgewater should invite Laws of the former and Huish of the latter railway company to a meeting to 'consider the subject of rates with a view to adjust them on a footing of equality to all parties'.[2] This was held, and seems to have led to a rates understanding between the three canal concerns and the Lancashire & Yorkshire. The Rochdale followed with a traffic arrangement also with the Manchester, Sheffield & Lincolnshire Railway, and the committee then told shareholders of their efforts at 'the allaying of that spirit of hostility and opposition which has been found so fatally injurious alike to Canal and Railway Interests'.[3] By this time the Aire & Calder were more worried about pressures nearer home, and had ceased to be so actively concerned with Manchester traffic, which they probably regarded as bound to decline.

The company adopted the Carriers' Act very soon after it was passed in 1845, but took no steps to start carrying because of their negotiations with the railway. Under an Act enabling them to vary tolls, they ordered traffic (except merchandise) passing not more than 3 miles* to pay maximum tolls on the grounds that a boat going and returning a short distance used as much water as one going farther, when water was a major item of expense.

At this time the company was providing carriers with a good deal of warehousing. Of their warehouses and large sheds, the following were in their own occupation: Sowerby Bridge, warehouse; Gauxholme, warehouse; Rochdale, 2 warehouses, 3 sheds; Heywood, warehouse, shed; Manchester, 3 warehouses. Wm Jackson & Sons had a shed and warehouse at Sowerby Bridge, 3 sheds and part of a warehouse at Rochdale, and one at Manchester; the Merchants Co 3 sheds and a warehouse at Rochdale, and one at Manchester; Kenworthy & Co, 2 sheds at Rochdale; J. & J. Veevers, a shed at Rochdale and a Manchester warehouse; Faulkner's and also Marsden's, a Manchester warehouse.

But at the end of 1845 Jackson's, a substantial firm with 32 wide and 6 narrow boats, and some 120 horses, fearing the canal would be sold to the railway, offered the business at about £12,000, and the company considered buying it. Early in 1846 the Jacksons wrote in to say that there was not sufficient margin between the freight rates they received, which had to be the same as the railway carriers, and the canal tolls to make it worth while to run boats. Veevers agreed. Subsequently Jackson seems to have worked on behalf of the railway company as a carrier on the canal. Later both became agents for the Bridgewater trustees.

Affairs went forward until 1855, when on 23 July agreement was reached for the Rochdale Canal to be leased for 21 years at an annual rent of £37,652 to four railways, all canal-owning, in the following

* Extended to 4 miles in June 1847.

proportions:* Lancashire & Yorkshire (73 per cent); Manchester, Sheffield & Lincolnshire (12½ per cent); North Eastern (8¼ per cent); London & North Western (6¼ per cent). The lease dated from 1 September, and provided that the canal company should pay all expenses of working. One member of each railway board was to join the committee. The rental was a sum sufficient to pay a dividend of £4 per share (£22,652 per annum) and provide £15,000 per annum for maintenance.

It was at once agreed that the railways might raise Rochdale Canal rates on salt, timber and grain if they wished, 'the Canal Company not offering or affording any facilities to the Aire & Calder or other Water Companies'.[4] It seems that the railway lessees now organized a department to carry by water between Manchester and Hull.

Here are the figures for the years between the opening of the Liverpool & Manchester in 1830 and the lease of the Rochdale Canal:

Years	*Tolls*	*Tonnages*	*Dividend per £85 share*			
	£	*Tons*	£	*s*	*d*	
1830–32	40,123	539,081	4			
1833–35	49,875	677,043	5	3	4	(£5·16½)
1836–38	59,174	828,926	7			
1839–41	53,838	839,640	6			
1842–44	26,901	751,269	3	16	8	(£3·83½)
1845–47	29,438	914,924	3	3	4	(£3·16½)
1848–50	24,542	769,316*	2	10	0	(£2·50)
1851–53	26,109	862,133	2	10	0	(£2·50)
1854–56	25,039	785,474	3	6	8	(£3·33½)

* Average of 1848–9; 1850 missing.

The thirties were a time of rapidly expanding trade on the canal, and both receipts and tonnages rose spectacularly, helped to some extent, perhaps, by the carriage of railway construction material. The opening of the Manchester & Leeds Railway caused a drop in tonnage from the highest figure so far reached, 875,436 for 1839, to a low of 667,311 in 1842, but a much more severe drop in toll receipts, from £62,712 to £27,266. Price cutting then brought traffic back to a new high of 979,443 in 1845 but toll receipts remained almost stationary at £28,695. In that year the dividend was £3 10s (£3·50) per £85 share, against the £6 of 1839 and the £4 of 1842. Thenceforward traffic and toll receipts kept pretty

* These were to apply to the first year, and could then be altered after the companies had analysed the traffic figures.

steady until the lease: in the last year of independent working, tonnage was 820,142, toll receipts £26,181, and the dividend £3. Given that total of toll receipts, to which should be added that for rents and other income, the new rental of £37,652 shows that the railways reckoned a substantial sum was worth paying to end competition. In fact, these lost £116,459 between 1 September 1855 and 31 December 1875, an average of £5,727 p.a. From their £15,000 p.a., the company were able to save sums which, with revenue from the occasional sale of lands and easements, enabled them to set up a reserve fund for improvements and the repayment of loans.

After the railway lease expired on 31 August 1876, the same arrangement was continued by mutual consent until the end of 1890, when it ended after an understanding that the railway companies still participating, the Lancashire & Yorkshire, Manchester, Sheffield & Lincolnshire and London & North Western, should pay £15,000 as a solatium. This was done in four half-yearly instalments. The £4 dividend lasted to the end of 1891.

Traffic and receipts rose during the lease; dividends remained steady, as the following figures show:

Years	*Tolls*	*Tonnages*	*Dividend per £85 share*
	£	*tons*	£
1857–59	23,048	754,421	4
1860–62	26,095	802,902	4
1863–65	22,040	731,141	4
1866–68	24,490	758,803	4
1869–71	25,073	766,156	4
1872–74	26,925	813,187	4
1875–77	28,579	878,651	4

The years were fairly uneventful. Piccadilly was widened, and in 1864 the lock under it was moved 24 ft to the east—on which exercise the company made a profit of £800. In 1857 the North Staffordshire Railway tried to lease the Bridgewater Canal, but the scheme fell through, and around 1864 the trustees were again considering a transfer to railway interests, but their scheme also dropped (see Chapter XIV). At the other end of the Rochdale's line the Calder & Hebble was leased by the Aire & Calder from 1 January 1865: the Rochdale expectantly noted that 'it was likely the locks . . . would be lengthened'[5] on the Calder & Hebble as a result, but work did not start until the 1880s, and was not carried far.

Meanwhile Hollingworth reservoir had become something of a pleasure resort. There were steamers on it in 1865—but not on Sundays. In November 1865, Mr Newall the lessee was told of 'the immoralities which it is stated take place in connection with the dancing stages at Hollingworth'.[6] Mr Newall indignantly replied in January that 'the immoralities referred to, if existing, had not occurred on any part of the Reservoir or the adjoining property leased to him by the Company'.[7] It seems an unlikely time of year for Pennine immoralities in any case.

In September 1872 the Bridgewater Canal was bought by a consortium of railway interests, and became the Bridgewater Navigation Company. In May 1873 the chairman of this company, Sir Edward Watkin, wrote that

> 'After the Conference with the Aire & Calder Directors on Friday, I feel now disposed to recommend the Board of the Bridgewater Navigation Company to take a lease, with an option of purchase, of the Rochdale Canal.'[8]

The suggestion was a £4 dividend in perpetuity, with an option to buy at £90 per £85 share, either when the current railway lease ended, or before if it could be terminated, possibly with Aire & Calder participation. The canal committee did not think the offer especially attractive, and declined it.

It may be interesting to look at the pattern of traffic on the canal in 1864, and again in 1875: the classification and revenue were as follows:

	Tonnage	
	1864	*1875*
	Tons	*Tons*
Merchandise, grain and sundries	217,926	336,660
Timber	57,379	88,224
Coal	291,936	284,190
Stone and lime	135,385	137,270
Salt	29,053	24,076

	Toll receipts	
	1864	*1875*
	£	£
Merchandise, grain and sundries	9,381	15,904
Timber	2,214	2,100
Coal	3,139	3,613
Stone and lime	4,924	3,776
Salt	753	583

In 1864 most of the merchandise, grain and sundries traffic was coming off the Bridgewater—112,189½ tons out of 217,926. Another 59,691 came off the Ashton, or originated at Manchester; 22,690 came on to the canal at Sowerby Bridge, and 18,227 tons at Rochdale. The timber nearly all came off the Bridgewater—50,470 tons out of 57,379, most of it probably from the Leigh branch. The coal, again, mostly came off the Bridgewater—207,340 out of 291,936, and was either used in Manchester or went on to the Ashton. Much of the rest was off the Ashton, while 24,837 tons originated on the canal between Rochdale and Manchester, and 12,206 tons came on to it at Sowerby Bridge, the last only moving short distances. About a quarter of the stone and lime traffic came off the Bridgewater—35,884 tons out of 135,385; 46,880 originated in Manchester or came off the Ashton—probably nearly all of it off the Peak Forest Canal; 20,727 tons originated at Rochdale; 19,276 at Gauxholme, and 12,550 tons at Sowerby Bridge. Curiously, quite a lot of this traffic went on to the Bridgewater—in June, 764 tons from Sowerby Bridge, 398½ tons from Gauxholme, and 1,269½ tons from Manchester and the Ashton, against 4,125 tons off the Bridgewater in the same month. The salt trade was almost entirely off the Bridgewater—presumably from the Trent & Mersey, and over two-thirds went the whole length of the canal to Sowerby Bridge—21,671 tons out of 29,053.

Out of 730,676 tons of traffic, therefore, 434,856 was coming off the Bridgewater, 50,393 from Sowerby Bridge, 156,721 from Manchester wharves or the Ashton—probably at least two-thirds from the latter—and 51,311 from the Rochdale. But, with the notable exception of salt, nearly all of it travelled only a short distance on the Rochdale, to Manchester wharves or the Ashton Canal.

For 1875 we can only quote figures for the month of June. In merchandise and grain, out of 25,579 tons, 16,024 is coming off the Bridgewater; 4,091 from Rochdale, 2,609 from Manchester and the Ashton, 1,547 from Sowerby Bridge. Nearly all the timber is off the Bridgewater, going to Manchester and the Ashton; most of the coal—15,686½ tons out of 21,974—is doing the same. A good deal of stone and lime is coming off the Bridgewater—4,634 tons, mainly from Manchester and the Ashton, but a similar quantity, 4,249½ tons, is coming from Manchester and the Ashton for the Bridgewater. The latter is probably limestone off the Peak Forest. The salt trade has now altered in character—only about one-third of what comes off the Bridgewater goes through to Sowerby Bridge—551½ tons out of 1,964 tons (the year's figures are 8,050 and 22,749).

In 1875 the railway lessees decided to give up carrying by water between Manchester and Hull. William Jackson & Sons thereupon bought the boats, helped by a loan of £1,800 from the Aire & Calder, to continue the trade.

In April 1880 a traffic arrangement was reached with the Leeds & Liverpool company, whereby the market should be shared, each to quote high rates in the other's area.

As part of the arrangements under which the Manchester Ship Canal Company took over the Bridgewater Canal, Castlefield lock, which had been Bridgewater property, became the responsibility of the Rochdale on 31 August 1887. The Ship Canal Act of 1885 also repealed the compensation tolls that the Bridgewater authorities were empowered to take under the Act of 1794.

In the late 1880s, it seemed as if carrying was about to fall away. The company therefore decided to form its own carrying department, buying the boats and equipment of one firm of carriers then, and others later, and issuing £48,000 worth of 3½ per cent debentures to finance it. They also bought two steamers in 1888, *Grace* and *Ceres*, and another, *Pioneer*, began to work in the following year. The decline was not, however, arrested, though it was probably slowed. In 1885, 33,468 boats, of which 6,780 were empty, used the canal, the average load being about 35 tons. In 1893 the figure was 26,098 (5,011). But, whereas in 1889 £2,289 was paid as tolls by the carrying department, in 1893 the figure was £6,978, or nearly one-third of the toll revenue. In one week in May 1893, 212 wide boats passed through Brownsfield lock, the first above the Ashton junction; 93 pairs of narrow boats, and 31 single narrow boats. Higher up, it is likely that the proportion of wide boats to the total would have been greater.

At the end of 1892 the carrying department possessed 15 steam cargo craft, 12 of steel and 3 of iron; 38 wooden narrow boats, 15 short wide boats able to work through the Calder & Hebble to Hull, and 42 boat horses and ponies. They also had 28 carts and dray horses to pull them. From 1896 some of these craft became surplus, and were disposed of.

By an Act of 1894, the original 5,663 shares of £85 each were converted into £481,355 of ordinary stock. By another of 1899, the capital was increased to £752,780 nominal by capitalizing the estimated sums that had from time to time been devoted to capital purposes out of income. Shareholders were then allotted £1 11s 3⅓d (£1·56) stock per £1 held as from 1 January 1900. The same Act granted the company the power to sell water, and changed its name from the Company of Proprietors of the Rochdale Canal to the Rochdale Canal Company.

In 1894, owing to a carrier giving up his business, there was a falling off in tonnage to and from east coast ports. However, the company arranged some through tolls with the Aire & Calder. The Calder & Hebble did not agree, but, the case having been referred to the Railway Commissioners, they accepted the Rochdale's case. In 1896 these through tolls were extended to traffic between the Bridgewater Canal and the Calder & Hebble.

At the beginning of 1904, a three-year market agreement came into force between the Manchester Ship Canal Co and the Rochdale Canal, with specially cheap tolls. In 1905 418,786 tons of traffic was interchanged with the Bridgewater, mainly to and from the ship canal.[9]

Carrying was given up at the end of World War I. During the time the company had run their own carrying department, they had slowly increased their percentage of traffic carried from about 20 per cent in the early 1900s to 30 per cent between 1910 and 1914. During the war, however, it fell back to 27 per cent.

When wartime control ended on 31 August 1920, the company could not immediately apply higher rates. Wages had gone up, hours been reduced, and weekend, Sunday and Saturday afternoon working was not now acceptable. The company's opinion that 'canal traffic cannot effectively be carried on except by continuous working', especially when boats were dependent on tide times, and the onset of depression, decided them to end carrying from 9 July 1921. However, as the 1922 report said, 'the discharged boatmen who hire our boats are doing a nice little business; they work all the hours they can'. But, all the same, it was a declining business. Two years later the report said: 'if ever there is to be a proper resuscitation of Inland Canals, it will have to be by Government affording adequate assistance, or taking them over and developing them'. About this time through use of the whole canal came to an end: in future shorter trips only were to be worked.

In 1923, the company had eight reservoirs, Blackstone Edge, Light Hazzles, Whiteholme, Warland, Hollingworth, Higher Chelburn, Lower Chelburn, and Easterly Gaddings Dam. The Oldham and Rochdale Corporations Water Act of that year[10] transferred the works and water rights of Blackstone Edge and Whiteholme to Oldham, Light Hazzles and Warland to Rochdale, and Hollingworth and the Chelburns jointly, with the water rights of Easterly Gaddings Dam. For these the company received £396,667, out of which they had to pay a quarter to the Manchester Ship Canal Co in consideration of the latter's loss of rights to take waste water from the Rochdale Canal; the company's net gain was £298,333. They also retained water rights in various streams, and a claim on reservoir supplies under certain conditions. They undertook to put their waterway into good order and reduce leakage to a minimum, to enable barges of 4 ft draught to navigate.

Certain of these monies were applied in buying its ordinary stock at prices not exceeding £50 per £100 nominal; £210,690 worth were bought at an average price of 49½, so reducing the stock in issue to £542,090.

Traffic declined steadily, and toll revenue with it, the last working trip being made over the full length of the canal in 1937. An Act of 1952[11] authorized the extinguishment of public rights of navigation over the whole canal except the section from a little above the Ashton Canal junction to the Bridgewater Canal, on the grounds that traffic had been negligible for many years. This exception was removed by an Act of 1965, the power not to apply, however, to the section from the Ashton junction to Castlefield until the owners of the Ashton Canal (the British Waterways Board) were authorized to close that. But interest in pleasure cruising was increasing, and the company could report for 1966: 'The mile length of canal in Manchester . . . is now being used for pleasure purposes. . . . To meet the promised increased use in the most beneficial way for the Company, a Lease of navigation rights over the uniting section has been offered to the pleasure boat organisation concerned.'

In its later years the company had to rely less and less on tolls. Instead, it derived a steadily increasing revenue from investments, the ownership of property, and the sale of water, so that for 1968 it had investments of a market value, with loans, of some £746,000, and paid a 7½ per cent dividend. In that year the company were looking forward to developing its land near the centre of Manchester with extensive offices and other commercial uses 'with an amenity focus based on the retained section of open urban waterway'. Transfer of parts of the canal to local authorities was reported, and 'interest in recreation facilities afforded by inland waterways, such as enjoyed by angling associations, is increasing and beginning to assist with canal maintenance to mutual advantage'.

CHAPTER XIX

The Remaining Lines

Manchester, Bolton & Bury Canal

THE Manchester, Bolton & Bury[1] remained a coal-carrying waterway of some importance until the 1930s, though most of the important supplying collieries were worked out by the 1900s.

Subsidence damage was extensive and costly. Between 1875 and 1888 part of the canal near Park House bridge fell 7 ft, on the Bolton–Prestolee length 6 ft 3 in., and between Bury and Prestolee 6 ft. After a serious burst at Agecroft in 1881, Sir John Hawkshaw reported to the Lancashire & Yorkshire Railway on the canal's condition, and as a result Edwin Muir engineered major reconstruction work. This went on until 1888, and included building long lengths of retaining wall on the side nearest the river, and re-puddling and concreting the bed.[2] The result was to cut maintenance costs and enable the canal to make a working profit:

	Coal & Coke tons	*Merchandise* tons	*Total Tonnage* tons	*Tolls* £	*Total Revenue* £
1888			620,345		8,069
1898	599,753	95,964	695,717	8,414	12,269
1905	567,345	86,804	654,149	7,503	11,884

Because of this subsidence, in 1884 the Lancashire & Yorkshire Railway began proceedings against a colliery firm, Knowles & Sons, for damages under the powers of the canal's 1791 Act, and after taking the case to the House of Lords, were successful. Thereupon settlements of other outstanding claims were negotiated by the railway company with a number of colliery concerns. A railway Act of 1892 gave them satisfactory amended powers to control the effects of mining under the canal, the company claiming that their 'action against Messrs. Knowles & Sons was the first in which the rights of canal companies under the old Canal Acts, with regard to support from minerals was decisively settled'.[3]

Much of the coal in the 1905 tonnage figure was carried to canalside works, often in 35 cwt containers which had been placed in the boats by steam cranes at the collieries; six mines were on the main canal and one on Fletcher's. Bricks, salt, manure, pulp and cotton were other substantial traffics. Most of the carrying was between points on the canal itself, only 124,233 out of 654,149 tons passing through the Irwell locks. By the turn of the century, sales of industrial water had begun to be an important source of revenue, though potential supplies had been lessened by water rights given to other authorities.[4]

Unlike the Great Central, the Lancashire & Yorkshire Railway was accused of discriminating against canal and in favour of rail-borne traffic by Mr F. Morton of Fellows, Morton & Clayton, the canal carriers, before the Royal Commission on Canals & Waterways.[5]

Thereafter, traffic and revenue steadily fell:

Year	*Tonnage*	*Receipts**	*Expenses*
		£	£
1913	467,653	8,209	4,774
1922	275,846	10,200	10,100
1928	149,993	7,518	8,854
1938	86,835	4,870	7,775
1946	30,368	7,296	12,500

* Only a small part from tolls: £1,084 in 1938, £471 in 1946. The major source of revenue was the sale of water.

Most of the tonnage was coal, and this was reduced by such colliery closures as Wet Earth on Fletcher's Canal in 1928. There were serious bursts in 1936, and in September 1939 the Minister of Transport ordered a ½ mile section at Clifton junction to be de-watered as a wartime precaution against flooding should there be bombing nearby.

Under a London, Midland & Scottish Railway Act of 1941, 6⅞ miles of the canal were abandoned. Two sections were left open, one of 4½ miles from Bury towards Prestolee, along which coal was carried from Ladyshore colliery, and another of 3½ miles between the southern end of Clifton aqueduct and the Irwell. The de-watered section, which had been piped, though not abandoned, was not reopened after the Minister's direction had been withdrawn in 1945; Ladyshore closed in 1951, and thereafter the only traffic at

Bury was the transport of coal across the canal for about 100 yd from the wharf to a canalside works. A British Transport Commission Act of 1961 abandoned the rest of the canal. Many of the canal works have since been filled in or demolished, but Clifton and Prestolee aqueducts still stand as major reminders of a once-important waterway.

Fletcher's Canal

In 1864 J. & J. Evans of Haydock leased the mining rights and canal. Two years later, with the Pilkington brothers, they formed the Clifton & Kearsley Coal Co. This firm was later absorbed by Bridgewater Collieries which in turn became part of Manchester Collieries. A rail wharf was provided beside the loading basin at Wet Earth. The tracks then ran parallel to the canal to Botany Bay and so to the Lancashire & Yorkshire at Clifton Junction.

Serious subsidence occurred in 1880, the canal being closed for seven months. In 1892 Botany Bay colliery closed. In 1905, however, the canal carried 142,905 tons of coal, and no other traffic. Of this total, 110,536 was loaded and discharged on the canal, i.e. carried from mine to railway, the balance of 32,369 being carried down the Manchester, Bolton & Bury. The revenue from tolls is given as £1,551. On the other hand, the Manchester, Bolton & Bury figures in the same set of Royal Commission statistics give 67,348 tons as the figure for coal leaving Fletcher's for that canal.

Wet Earth colliery closed in 1928, after which Fletcher's Canal was no longer used to carry coal. The only traffic was now felspar and clay for a short distance inwards to the Pilkington Tile Co's works near the canal entrance. Fletcher's Canal became disused in 1935.[6]

Ashton-under-Lyne Canal

The Manchester, Sheffield & Lincolnshire Railway, later the Great Central, worked the three canals it owned, the Ashton, Peak Forest and Macclesfield, as a single concern, operating a small-scale carrying department over all three, mainly from a rail/canal interchange point at Princes Dock, Guide Bridge. There were also interchange facilities at Ashton, where the canal ran by the railway's goods warehouse, and at Manchester London Road. Coal from the New Moss colliery came on to the main line at China Bridge. Other traffic came from chemical works in the Clayton area, cotton mills at Ashton and Oldham, locomotive works and sheds at Gorton, and collieries on the Fairbottom and Hollinwood branches.

The railway company ceased to carry on their three canals in 1892. Till then boats had been run to a timetable in connection with goods train services at Guide Bridge station, and so had enabled the owning company to penetrate London & North Western and Midland territory. Sir Sam Fay in 1906 said that the service ended 'because whilst we got traffic in one direction for our boats we had to bring them back empty'.[7] In 1883 the canal companies, which had been kept in existence to receive and distribute the railway's annual payments, were dissolved under the Manchester, Sheffield & Lincolnshire Railway (Additional Powers) Act of 2 August, the shareholders being given railway stock for their shares.

In 1858 the canal carried 433,526 tons, and had a revenue of £9,370.[8] By 1905 these had fallen to 241,176 tons and £8,202, with a loss on the year of £2,000, though H. R. de Salis told the Royal Commission in 1906 that he considered that both the Ashton and the Peak Forest were maintained as well as the average independent canal,[9] and F. Morton of Fellows, Morton & Clayton testified to fair treatment from the Great Central Railway.[10] Traffics were still very varied in 1905, the most important being:[11]

	tons		*tons*
Coal	102,552	Tar	17,808
Cotton	31,368	Lime and limestone	13,482
Grain	22,572	Acid	11,676

About 1900 a steam passenger boat was sometimes run from Bardsley bridge on the Fairbottom branch to Crime lake on the Hollinwood line and beyond.[12]

Thereafter the growth of short-distance road transport had a disastrous effect on the canal. The Hollinwood branch was fairly well used until about 1928, when traffic began to fall away. Two leakages on the Fairbottom branch in 1930 made it necessary to lower the water level, and colliery subsidence caused the Hollinwood branch to be closed early in 1932. The upper portion of the Hollinwood branch for 3½ miles was abandoned by the British Transport Commission in 1955, the rest in 1961. The Stockport branch ceased to carry commercial traffic about the early 1930s, and was abandoned by the Commission in 1962. The Ashton's main line ceased to be appreciably used commercially after the 1930s, in 1947 carrying only 5,452 tons, and became unnavigable throughout about 1960. Such revenue as the canal continued to earn came

NORTH STAFFORDSHIRE

RAILWAY COMPANY,

TRENT & MERSEY NAVIGATION BRANCH,

CARRIERS,

BY

CANAL AND RAILWAY,

TO AND FROM

LONDON,
BIRMINGHAM,
WOLVERHAMPTON,
STOURBRIDGE,
DUDLEY and through the
South of STAFFORDSHIRE
SOUTH WALES.

BRISTOL,
CHELTENHAM,
GLOUCESTER,
The POTTERIES,
MACCLESFIELD,
CONGLETON,
LEEK, &c. &c.

KIDDERMINSTER and all Parts of the West.

NOTTINGHAM,
DERBY,

BURTON,
LICHFIELD.

OFFICES IN MANCHESTER,

No. 44, GEORGE-STREET.

WHARF,

ASHTON CANAL, DUCIE-STREET,

PICCADILLY.

PRINCIPAL ESTABLISHMENTS.

LIVERPOOOL.........Old Quay Dock.
WOLVERHAMPTON..Albion Wharf.
BIRMINGHAM ....Great Charles-street.
BRISTOL ....Quay Head.

LONDON,

16, Wharf Road, City Basin---Receiving House,
White Horse, Cripplegate, and London Wall.

33. The boat services of the North Staffordshire Railway operate from the Ashton Canal wharf. An advertisement of 1848

mainly from the sale of water. It is not yet known whether the canal will be restored for pleasure use, or will be abandoned.

Peak Forest Canal

Under the ownership of the Manchester, Sheffield & Lincolnshire Railway the canal remained busy. In 1838, under company management and before the through trade by way of the Macclesfield Canal had been greatly affected by railway competition, the tonnage carried was 442,253½. In 1848, under railway control, this had fallen to 343,549, but by 1858 had recovered to 378,889½, though receipts per ton carried had fallen from 10·4d in 1838 to 7·6d in 1848 and 5·4d in 1858.

In 1845 the Stockport, Disley & Whaley Bridge Railway was incorporated, under the London & North Western's wing, to connect with the Cromford & High Peak Railway, after the Manchester, Sheffield & Lincolnshire had sought assurances that canal traffic would as far as possible be protected.[13] The line was opened in June 1857 and absorbed by the London & North Western in 1866.

John Boulton, again using former Glasgow, Paisley & Ardrossan Canal swift boats, probably in the early 1840s ran summer excursion boats and started a packet service, first from Macclesfield, later from the bottom of the locks at Marple past Hatherlow (where Stockport Road, Romiley, crosses the canal), Woodley, Apethorn and Hyde, first to Dog Lane station, Dukinfield, on the Manchester, Sheffield & Lincolnshire Railway's main line, and then to Dukinfield station on that company's branch from Guide Bridge to Stalybridge after it opened on 23 December 1845.[14] He also built and ran the Queen's Hotel at Marple. Probably from the beginning, and certainly in 1851, there were one weekday boat from Marple and two more from Hatherlow, with four boats each Sunday through from Marple. The fare was 1s (5p) from Marple to Dukinfield first cabin, and 8d second. After the Dukinfield to Hyde branch of the Manchester, Sheffield & Lincolnshire Railway had been opened on 1 March 1858, the timing of the services was slightly altered so that boats connected with trains at either Dukinfield or Hyde.[15]

Later, the boats seem to have become less efficient. A local resident, looking back from 1881, wrote that the boat

'was so swift that I have scores of times started from Ashton at the times announced as its starting time, and have arrived at

Marple, by walking before it. It was not that its pace was slow when it was going, but at every station the captain had to go to the public-houses to hunt up the passengers, and then the passengers had to go and hunt up the captain. The owner, a well-known, clever Ashton man, took great pride in keeping his boats nice'.

One was called the *Harlequin*.[16]

In 1883 the canal company was dissolved and the canal and tramroad vested in the railway company. By the end of the century it had greatly declined from its heyday. Though in his evidence to the Royal Commission on Canals and Waterways de Salis described it as 'quite up to the average' of the independent canals in its standard of maintenance, it only carried 136,148 tons in 1905, at an average toll of 4·6d per ton. Of this total, 15,660 tons was lime and limestone, other major traffics being coal, 37,446 tons, stone, 27,642 tons, cotton, 18,942 tons, and grain, 16,206 tons. Revenue in that year was £4,138, and the operating loss £976. Subsidence was then affecting the line at Hyde and Dukinfield.

The last known train of waggons on the tramroad came down in 1922, and the line was abandoned in 1925.[17] When it was closed, the upper level had little to do. On the rest of the canal some through trade continued with the Macclesfield Canal and some coal and raw materials moved on the lower level towards the Ashton. Today, the upper level is popular for cruisers from the Macclesfield, while the locks and the lower section are in a similar position to the Ashton.

Huddersfield Canal

About the time that the canal was leased to the London & North Western Railway, Pickford's seem still to have been carrying—it must have been one of the last of their canal services. A directory of 1848[18] says that by their boats: 'Goods are forwarded by steam or sailing vessel to London, Yarmouth, Berwick, Edinburgh, Leith, Dundee, Aberdeen and other parts of the East Coast.'

Those who used the tunnel saw many vicissitudes:[19] on 23 April 1855 a boat owned by Thomas Nicholson carrying stone window sills from Marsden to Manchester capsized when the load moved. The leggers were three hours in the water before help arrived, and the tunnel had to be drained to move the boat and its cargo. On 21 February 1857 the tunnel was closed for several weeks for repairs as dangerous and unsafe. In November 1869 a 3 ton rock, said to have been dislodged by blasting for the railway tunnel, fell on to a Manchester, Sheffield & Lincolnshire Railway Co's boat when it was about 100 yd from the Marsden end, and sank it. In December a heavy fall of rock reduced the depth at one point to 18 in. In 1870, while a railway tunnel was being built, a small boat carrying gunpowder to the railway workings blew up, killing both boatmen and putting out all the workmen's lights.

The canal tunnel had to be closed for part of the time that the double-track railway tunnel was being cut: from 4 September to 19 December 1892; 20 March to 10 April 1893, and again from 10 June when the railway company announced that it would be closed for six months, and that during that time the following traffics would be carried by rail: grain, Huddersfield to Stalybridge for Buckley & Newton; acid, Huddersfield to Greenfield for Holliday & Sons; bricks and undressed stone from Linthwaite to Mossley for the executors of John Walker.[20] In January 1894 it was stated that tunnel work would last a further six months.[44] The tunnel was now extended some 240 yd at the Diggle end, which bears the date 1893. Reopening was from 5 September 1894. There were finally two single and one double line railway tunnels.

So far, traffic had held up well, though it was deserting the summit and moving towards the ends. In 1848, 169,487 tons were carried, and in 1888, 179,570, these last yielding tolls of only £3,391 out of a gross revenue of £7,786, or 5·6d per ton. Ten years later traffic was 161,899 tons, but tolls had fallen to £1,935.

In June 1894 the London & North Western Railway appeared before a joint committee of both Houses against the charges imposed by the Board of Trade under the Railway & Canal Traffic Act, 1888, which, they said, had reduced their income by 43 per cent. In 1891, they told the committee, their revenue had been £3,441 and the expense of working the canal £4,300, the most expensive part to maintain, the summit, being the least used. The appeal failed, though the company were allowed to make a 2s 6d (12½p) minimum charge. In 1905, when traffic over the summit had ceased and most of what remained was concentrated at the western end, tonnages had fallen sharply to 97,939, and tolls to £1,015.[22]

Two other events of the late nineteenth century occurred when in 1875 the London & North Western legalized the sale of water from the canal after stormy opposition from Manchester corporation, and in 1882–3, when the canal was diverted for ¼ mile near Scout because of railway construction.

Traffic was seriously reduced during World War I, and disappeared during the second:

Year	*Tonnage*	*Receipts*	*Expenditure*
	tons	£	£
1913	91,753	3,382	4,059
1922	33,307	4,866	8,851
1930	28,006	5,358	7,655
1938	16,650	5,054*	7,404
1946	–	3,717*	10,270

* Toll revenue was £287 in 1938, nil in 1946.

In 1944 the canal was abandoned by a London, Midland & Scottish Railway Act, except for the ½ mile section at Huddersfield up to the third lock at Engine bridge. This was transferred to the Calder & Hebble company, but in turn was abandoned back to Aspley in 1963. The last boat through the whole length of canal and tunnel was probably L. T. C. Rolt's. He describes the experience in his *The Inland Waterways of England.*

The canal is disused, but the great tunnel is maintained to drain the railway tunnels and help provide industrial water, and an occasional boat trip through it still takes place.

Authors' Notes and Acknowledgements

OUR thanks are due to very many who have helped us.

To the Archivist and staff of the British Railways Board Historical Records for access to minute books and other records, without which this history could not have been written, and to the House of Lords Records Office for access to parliamentary records.

To Mr Alfred Gaskell, Mr Anthony Hall, Mr D. E. M. Hawkins, Mr A. Hayman, Mr Brian Lamb, Mr C. M. Marsh and Mr V. I. Tomlinson for kindly reading chapters in draft.

To Mr A. Hayman, manager of the Bridgewater Department of the Manchester Ship Canal Company and Mr D. E. M. Hawkins, manager of the Rochdale Canal Company, for so kindly making records available, and to Mr F. C. Mather for lending the manuscript of his forthcoming book, *After the Canal Duke*.

To Mrs J. H. Home, Mrs P. Paget-Tomlinson, Mr T. Walsh and the late Mr J. Clarke for reading local newspaper files.

To Messrs H. E. Cardy, R. H. J. Cotton, J. E. Freeman, O. H. Grafton, C. Hadlow, R. J. Hutchings and other members of the staff of the British Waterways Board; Mr J. ApThomas, Mr D. Ashcroft, Mr M. F. Barbey, Professor T. C. Barker, Mr G. H. Brown, Dr W. H. Chaloner, Mr N. Fraser, Mr J. S. Gavan, Mr R. Hargreaves, Dr J. R. Harris, Mr G. O. Holt, Mr K. Hoole, Mr J. Marshall, Mr I. P. Moss, Mr F. Mullineux, Mr P. Norton, Mr M. M. Schofield, Mr W. Skillern, Mr W. N. Slatcher, Mr J. Unsworth and Mr H. Tyldesley of Bridgewater Estates Ltd, Mr G. B. Vaughton of the Manchester Ship Canal Company, Mr A. P. Voce, Mr J. E. Weatherill and Mr Paul N. Wilson for all their help.

To the archivists and staffs of the Cheshire, Cumberland, Westmorland & Carlisle, Lancashire and Liverpool Record Offices, the librarians and staffs of Chetham's and the John Rylands libraries, the many public librarians and their staffs who have so frequently helped us, and especially Mr T. Walsh and the staff of the Local History Library of Manchester Central Library, who have been endlessly generous of their time and trouble.

To Mr Richard Dean for drawing the maps, and to Messrs E. & R. E. Pye Ltd, photographers, of Clitheroe, for the trouble they have taken in copying pictures.

Our thanks are also due to the following for permission to reproduce photographs and other illustrations: Manchester Public Library, plates on pages 49, 50, 286 (above), 304; Figs 1, 17, 23, 25, 33; R. J. Hutchings, British Waterways Board, 67 (above); Waterways Museum, 67 (below), 285 (above); St Helens Public Library, 69 (above); Professor T. C. Barker, 69 (below); G. J. Biddle Collection, 133, 134 (above), 151 (above), 152 (above), 388 (above); G. Ingle, 134 (below); Colne Public Library, 151 (below); Lancaster City Museum, 152 (below); R. Russell Craps, 285 (below); Accrington Public Library, 286 (below), 387 (above), Fig 11; D. G. Williams, 303; R. Brook, 369 (above); W. B. Stocks, 369 (below); Charles Hadfield, 370, 387 (below); Kendal Public Library, 388 (below); Lancashire C.R.O., Figs 3 and 4; Cumberland & Westmorland Joint R.O., Figs 12 and 13; John Rylands Library, Manchester, Figs 19 and 21.

The pictures on page 49 are taken from J. Corbett, *The River Irwell*, 1907. Fig 20 is taken from a plan made by Mr Brian Lamb, Fig 24 is from the *Illustrated London News* of 18 October 1851, and Figs 27 and 28 from *The Engineer* of 24 July 1908. Fig 32 is copied from the original owned by 'Ann Tique', North Road, Lancaster.

NOTES

Notes to Chapter IX

1. For a fuller account of the canal and the industries that grew up round it, to which we are indebted, see V. I. Tomlinson, 'The Manchester, Bolton and Bury Canal Navigation and Railway Company, 1790–1845', *Trans. Lancs & Cheshire Antiq. Soc.*, LXXV–VI, 1965–6.
2. John Rennie's MS *Notebooks* (I.C.E. Library).
3. Manchester, Bolton & Bury Canal Proprietors' Minute Book, 3 November 1790.
4. For Heathcote's connection with Gresley's Canal and the Newcastle-under-Lyme Junction Canal, see Charles Hadfield, *The Canals of the West Midlands*, 2nd ed., 1969.
5. 31 Geo III *c.* 68.
6. Manchester, Bolton & Bury Canal Proprietors' Minute Book, 25 June 1794.
7. S. R. Harris, 'Liverpool Canal Controversies, 1769–1772', *Journal of Transport History*, May 1956.
8. *Manchester Mercury*, 13 September 1791.
9. William Bennet also surveyed the Haslingden Canal. In the south-west he was engineer of the Dorset & Somerset Canal and concerned with the Ivelchester & Langport Navigation, for which, see Charles Hadfield, *The Canals of South West England*, 1967.
10. *Reasons*. Single-sheet broadside, n.d. (authors' collection).
11. J.H.C., 26 March 1794.
12. Manchester, Bolton & Bury Canal Proprietors' Minute Book, 6 May 1809.
13. Ibid., Committee Minute Book, 11 May 1796.
14. Tomlinson, 'Manchester, Bolton & Bury Canal & Railway', op. cit.
15. *Manchester Mercury*, 27 August 1799, 2 September 1800. *Cowdroy's Manchester Gazette*, 12 September 1801.
16. Manchester, Bolton & Bury Canal Proprietors' Minute Book, 27 June 1799.
17. Mersey & Irwell Navigation Minute Book, 7 June 1809.
18. 45 Geo III *c.* 4.
19. A plan of the Leigh branch in the Thomas Moore papers at the Lancashire Record Office shows the proposed Red Moss line. A connection then would have required 7 rising locks on a branch from the Leeds & Liverpool at the top of Wigan locks to Red Moss, and 11 falling locks thence to Bolton.
20. See Lois Basnett, 'The History of the Bolton and Leigh Railway based on the Hulton Papers (1824–1828)', *Trans. Lancashire & Cheshire Antiq. Soc.*, LXII, 1950–1.
21. For a full description of the collieries served, and their connection with the canal, see Tomlinson, 'Manchester, Bolton & Bury Canal & Railway', op. cit.
22. Pigot and Dean's *Manchester & Salford Directory*, 1815.
23. See Charles Hadfield, *British Canals*, 4th ed., 1969, p. 173. There is a different account in Manchester Central Library M/CR f. 1820/17.
24. Mrs Louisa Potter (1806–98), *Lancashire Memories*.
25. Pigot & Sons' *Manchester Directory*, 1836.
26. *Bolton Chronicle*, 31 July 1830.
27. Manchester, Bolton & Bury Canal Proprietors' Minute Book, 14 September 1830.

28. Ibid., 30 September 1830.
29. For further information on this scheme, see F. C. Mather, *After the Canal Duke*, 1970, Chapter IV.
30. 1 & 2 Will IV *c*. 60.
31. 2 & 3 Will IV *c*. 30.
32. Manchester, Bolton & Bury Canal Proprietors' Minute Book, 3 December 1832.
33. Peak Forest Canal Committee Minute Book, 18 August 1831.
34. Manchester, Bolton & Bury Canal Proprietors' Minute Book, 30 June 1836.
35. Peak Forest Canal Committee Minute Book, 30 January 1839.
36. See Tomlinson, 'Manchester, Bolton & Bury Canal & Railway', op. cit.
37. Mersey & Irwell Navigation Minute Book, 1, 9 March 1843.
38. 9 & 10 Vic *c*. 378.
39. Leeds & Liverpool Canal Committee Minute Book, 8 July 1793.
40. Ibid., 16 August 1793.
41. Aikin, *Manchester*, op. cit.
42. 34 Geo III *c*. 77.
43. Ibid.
44. *Manchester Mercury*, 15 July 1794.
45. See the Manchester, Bolton & Bury Canal papers in the Lancashire Record Office.
46. *Manchester Mercury*, 6 June 1797.
47. For tub-boat canals with inclined planes, see for instance Chapter IX of Charles Hadfield, *The Canals of the West Midlands*, on those in east Shropshire, or Chapter X of Charles Hadfield, *The Canals of South West England*, on the Bude Canal.
48. See A. G. Banks and R. B. Schofield, *Brindley at Wet Earth Colliery*, 1968.
49. Information from Mr Alfred Gaskell. Arthur Freeling's *Manchester and Bolton Railway Companion*, describing the first railway journey from Manchester to Bolton, mentions the level at Botany Bay. This is confirmed by oral evidence given to Mr Trickett of Walkden Records Office, where an old plan of the colliery appears to show the canal. Apart from the plan mentioned in the text, J. Corbett's *The River Irwell* alludes to the Wet Earth level, though he says it was 8 miles long, which was almost certainly not so.
50. This account of Fletcher's Canal has been written from Alfred Gaskell's papers *Th'Owd Cut* (1961) and *The Fletchers Canal* (1963) and from information given by the author; material in the Swinton & Pendlebury Public Libraries; the Minute Books of the Manchester, Bolton & Bury Canal; and the *Report* of the Royal Commission on Canals & Waterways, 1909.

Notes to Chapter X

1. *The Todmorden & Hebden Bridge Historical Almanack*, 1874.
2. For the Calder & Hebble Navigation, see Charles Hadfield's forthcoming book *The Canals of Yorkshire and the North East*.
3. Copy of subscription deed in Rochdale Canal records. See also *Williamson's Advertiser & Mercantile Gazette* (Liverpool), 29 August, 7 November, 26 December 1766.
4. John Rennie's MS *Notebooks* (I.C.E. Library).
5. *Williamson's Advertiser* (Liverpool), 26 December 1766.
6. For the Selby Canal, part of the Aire & Calder Navigation system, see Charles Hadfield's forthcoming book, *The Canals of Yorkshire and the North East*.
7. *Manchester Mercury*, March, April 1788.
8. Rochdale Canal Minute Book, 17 February 1791.
9. Ibid., 23 September 1791.
10. Rennie, *Notebooks*, op. cit.

11. Rennie, *Notebooks*, op. cit.
12. Rochdale Canal Minute Book, 23 August 1792.
13. Rennie, *Notebooks*, op. cit.
14. Rochdale Canal Minute Book, 4 October 1792.
15. *Case as to the compensation of the Duke of Bridgewater for the loss of his Warehouse-duties and Wharfage*, n.d. (I.C.E. Library).
16. *Aris's Birmingham Gazette*, 4 February 1793.
17. Rochdale Canal Minute Book, 30 December 1793.
18. *Manchester Mercury*, 18 February 1794.
19. 34 Geo III *c*. 78.
20. Report in Rochdale Canal Company's records.
21. First Report, May 1796.
22. Rochdale Canal Minute Book, 4 December 1795.
23. J. Phillips, *A General History of Inland Navigation*, 5th ed., 1805, p. 591.
24. *Manchester Mercury*, 19 December 1797, advertisement by Richard Milnes
25. Rochdale Canal Minute Book, 18 December 1800.
26. Ibid., 7 May 1801.
27. 40 Geo III *c*. 36.
28. *Derby Mercury*, 7 October 1802.
29. 44 Geo III *c*. 9.
30. Mersey & Irwell papers, Liverpool corporation (Liverpool P.L.).
31. *Morning Chronicle*, Thursday, 27 December 1804.
32. *Report* of 1805.
33. 40 Geo III *c*. 20.
34. Rochdale Canal Minute Book, 25 October 1805.
35. Ibid., 15 May 1806.
36. 47 Geo III *c*. 81.
37. Rochdale Canal Minute Book, 10 September 1806.
38. Ibid., 23 November 1810.
39. Pigot and Dean's *Manchester and Salford Directory*, 1815.
40. Mersey & Irwell Navigation Minute Book, 4 February 1807.
41. Rochdale Canal Minute Book, 2 May 1816.
42. Ibid., 25 October 1816.
43. Ibid., 21 November 1822.
44. Ibid., 16 January 1823.
45. Ibid., 24 April 1823.
46. Ibid., 4 February 1824.
47. Ibid., 8 August 1832.
48. Ibid., 4 November 1830.
49. Ibid., 14 March 1831.
50. Ibid., 7 August 1834.
51. Ibid., 8 February 1837.
52. Manchester & Leeds Railway Minute Book, 14 September 1840.
53. Ibid., 21 September 1841.
54. Ibid., 12 October 1840.
55. Ibid., 28 December 1840.
56. Ibid., 17, 24 May 1841.
57. Rochdale Canal Minute Book, 10 March 1841.
58. Ibid., 21 April 1841.
59. *Derby Mercury*, 30 November 1842.
60. Rochdale Canal Minute Book, 8 February 1843.
61. Ibid., 28 February 1843.
62. Ibid., 13 March 1844.
63. Ibid., 11 December 1844.
64. Ibid., 12 March 1845.
65. Ibid., 22 March 1845.
66. Ibid., 13 August 1845.
67. Ibid., 8 October 1845.

Notes to Chapter XI

1. For Sir John Ramsden's Canal and the Calder & Hebble, see Charles Hadfield's forthcoming book, *The Canals of Yorkshire and the North East.*
2. For the Macclesfield Canal, see Charles Hadfield, *The Canals of the West Midlands*, 2nd ed., 1969.
3. For the Cromford Canal, see Charles Hadfield, *The Canals of the East Midlands*, 2nd ed., 1970.
4. *Manchester Mercury*, 30 August 1791.
5. I am indebted to Mr J. A. Hall for much help in writing this account of the Ashton Canal.
6. *Manchester Mercury*, 13 September 1791.
7. Ibid., 18 October, 1 November 1791.
8. Ibid., 22 November 1791.
9. 32 Geo III *c.* 84.
10. J. Aikin, *Description of the Country from Twenty to Thirty Miles round Manchester*, 1795.
11. 33 Geo III *c.* 21.
12. For the 1790 project, see the *Manchester Mercury*, 16, 23 November 1790.
13. *Derby Mercury*, 26 September 1793.
14. Ashton-under-Lyne Canal Minute Book, 8 August 1798.
15. A. Rees, *Cyclopaedia*, art. 'Canal', 1819.
16. Rochdale Canal Minute Book, 4 December 1795.
17. *Manchester Mercury*, 5 September 1797.
18. For this episode, see the Hulton papers (Lancashire C.R.O. DDHu 32/4).
19. 38 Geo III *c.* 32.
20. Huddersfield Canal Committee Minute Book, 16 February 1798.
21. Ibid., 22 March 1798.
22. Ashton-under-Lyne Canal Minute Book, 8 August 1798.
23. Huddersfield Canal Committee Minute Book, 18 December 1800.
24. Ashton-under-Lyne Canal Minute Book, 22 October 1804.
25. There was an older Newcomen engine, 'Fairbottom Bobs', which dated from 1760. Before the canal came, this had been used for colliery drainage, being assisted by waterwheels driven by the water it was pumping out. After the canal had been built, this engine supplied it until 1830. See also L. T. C. Rolt, *Thomas Newcomen*, 1963, p. 139.
26. 39 & 40 Geo III *c.* 24.
27. 45 Geo III *c.* 11.
28. In 1842 there was a debt of £12,000 and 1,766 shares of an average value of £97 18s (£97·90). This gives a figure of £184,891 at that date.
29. *Hereford Journal*, 17 December 1806.
30. *Manchester Mercury*, 27 June 1797.
31. Ibid., 2 July 1799.
32. Ashton-under-Lyne Canal Minute Book, 28 May 1802. The road was to be Store Street. Progress must have been slow, for Joseph Aston, *A Picture of Manchester*, 1816, described it as 'intended'.
33. Peak Forest Canal Minute Book, 7 March 1808.
34. B. Baxter, *Stone Blocks and Iron Rails*, 1966, p. 164.
35. Huddersfield Canal Committee Minute Book, 4 February, 9 March, 27 April 1825.
36. Peak Forest Canal Minute Book, 30 January 1839.
37. George Dow, *Great Central*, Vol. I, p. 42.
38. A. R. Bennett, *The Chronicles of Boulton's Siding*, 1927, pp. 14–15.
39. 11 & 12 Vic *c.* 86.
40. 34 Geo III *c.* 26.
41. *Manchester Mercury*, 30 April, 14 May, 18 June 1793.

42. For Oldknow, see G. Unwin, A. Hulme and G. Taylor, *Samuel Oldknow and the Arkwrights*, 1924.
43. Op. cit., pp. 156–8.
44. Mr Brian Lamb tells me that Rosehill tunnel is shown on the 1849 tithe commutation map for Romiley.
45. George Borrow, *Wild Wales*, 1862 (Everyman ed., p. 40).
46. Peak Forest Canal Committee Minute Book, 11 September 1794.
47. Ibid., 4 March 1795.
48. Gloucester & Berkeley Canal Committee Minute Book, 3 February 1795.
49. Peak Forest Canal Committee Minute Book, 22 April 1795.
50. Ibid., 23 April 1795. For the canals of the Coalbrookdale area, see Charles Hadfield, *The Canals of the West Midlands*, 2nd ed., 1969, Chapter IX.
51. Ibid., 18 May 1795.
52. Ibid., 3 June 1795.
53. Ibid., 11 September 1795.
54. Ibid., 2 March 1796. The engineer to have built six 20–25 ton 'trading boats' and 'so many small Boats to be made as with those already made and making will make ten'. But these may have been construction boats.
55. See Charles Hadfield, *The Canals of the West Midlands*, 2nd ed., 1969, p. 210, and *Manchester Mercury*, 16 February, 29 March, 26 April, 24 May 1796.
56. *Derby Mercury*, 8 September 1796.
57. 'Samuel Oldknow and the Marple Lime Kilns' (Manchester Central Library F.666/9/OLI).
58. A. Rees, *Cyclopaedia*, art. 'Canal', 1819.
59. 39 & 40 Geo III *c.* 38.
60. See D. L. Frank, 'The Peak Forest Tramroad', *Railway Magazine*, 1941, pp. 337–9; B. Baxter, 'Early Railways of Derbyshire', *Engineering*, 17 June 1949, and B. Baxter, *Stone Blocks and Iron Rails*, 1966.
61. Mr Brian Lamb has done a great deal of research on the industrial archaeology of the canal, Bugsworth and the tramroad, and has written up his results. We are indebted to him for much information.
62. *Derby Mercury*, 27 November 1800.
63. He leased his kilns at Marple to others in 1811, and after that did not take an active part in the canal's affairs. He died in 1828.
64. The locks seem originally to have been fitted with some sort of compressed-air apparatus to work the paddles, but it failed, and was almost at once replaced by normal paddle gear.
65. Information from Mr Brian Lamb.
66. 45 Geo III *c.* 12.
67. Information from Mr Brian Lamb.
68. Peak Forest Canal Committee Minute Book, 19 November 1823.
69. Ibid., 23 February 1825.
70. 2nd ed., 1969, pp. 212–14.
71. Peak Forest Canal Committee Minute Book, 14 October 1841.
72. Ibid., 26 April 1843.
73. BTHR. File of Manchester, Sheffield & Lincolnshire Railway reports.
74. Information from Mr J. A. Hall and Mr Brian Lamb. This navigation is not included in Appendix I.
75. For the Don Navigation, see Charles Hadfield's forthcoming book, *The Canals of Yorkshire and the North East*
76. For the Chesterfield and Cromford canals, see Charles Hadfield, *The Canals of the East Midlands*, 2nd ed., 1970.
77. Cromford Canal Proprietors' Minute Book, 25 November 1802.
78. Peak Forest Canal Committee Minute Book, 25 May 1803.
79. For the Grand Union Canal, Leicester line, and Erewash Canal, see Charles Hadfield, *The Canals of the East Midlands*, 2nd ed., 1970.
80. Grand Junction Canal Committee Minute Book, 10 April 1810; Grand Union Canal Minute Book, 13 December 1810.

81. Peak Forest Canal Committee Minute Book, 11 June 1810.
82. Cromford Canal Proprietors' Minute Book, 19 September 1810.
83. Rennie's MS *Notebooks* (Institution of Civil Engineers' Library), and *A Plan of the proposed High Peak Forest Junction Canal* (British Museum).
84. Cromford Canal Proprietors' Minute Book, 11 December 1810.
85. Peak Forest Canal Committee Minute Book, 24 August 1813.
86. George Dow, *Great Central*, 1960, Vol. I, p. 1.
87. Leicestershire & Northamptonshire Union Canal Minute Book, 19 November 1810; *Aris's Birmingham Gazette*, 25 February 1811.
88. *Derby Mercury*, 1 and 8 November 1810.
89. Ibid., 13 December 1810.
90. Cromford Canal Proprietors' Minute Book, 29 May 1811.
91. Nottingham Canal Minute Book, 3 June 1811.
92. *Derby Mercury*, 22 September 1814.
93. *Prospectus* of the Grand Commercial Canal (Waterways Museum).
94. Joseph Haslehurst, *Second Report on the proposed Grand Commercial Canal*, 1824 (I.C.E. Library).
95. *Derby Mercury*, 21 July 1824.
96. Ibid., 25 August 1824.
97. Ibid., 3 November 1824.
98. Ibid., 17 November 1824.
99. Ibid., 15 December 1824.
100. 6 Geo IV *c*. 30. For the origin and early history of the railway, see D. J. Hodgkins, 'The Origins and Independent Years of the Cromford & High Peak Railway', *Journal of Transport History*, May 1963.
101. German Wheatcroft was a considerable carrier on the Nottingham and Cromford canals, the Cromford & High Peak Railway and the Peak Forest Canal and tramroad, and had his own horses and carts in Manchester.
102. Nottingham Canal Minute Book, 20 March, 10 May 1832.
103. Peak Forest Canal Committee Minute Book, 7 December 1843, 22 January 1844.
104. Ibid., 11 November 1825.
105. *Derby Mercury*, 24 January 1827.
106. Peak Forest Canal Committee Minute Book, 18 September 1828.

Notes to Chapter XII

1. For Sir John Ramsden's Canal and the Calder & Hebble Navigation, see Charles Hadfield's forthcoming book, *The Canals of Yorkshire and the North East*.
2. William Pontey, *A Short Account of the Huddersfield Canal*, n.d. (*c*. 1812) (I.C.E. Library).
3. Huddersfield Canal *Report*, 23 October 1793 (BTHR HRP 6/8).
4. *Manchester Mercury*, 1 April 1794.
5. 34 Geo III *c*. 53.
6. We are grateful to Mr R. H. J. Cotton, Assistant Chief Engineer (Operations) of the British Waterways Board, for the scaled length quoted. For the alterations, see 'The "London & North-Western" Tunnels', *L.M.S.* Magazine, March 1925. The original length of 5,456 yd given in *Bradshaw's Canals and Navigable Rivers*, 1904 edition, seems to be correct and not the 5,415 yd of later editions.
7. Huddersfield Canal Committee Minute Book, 29 September 1796.
8. Ibid., 15 September 1797.
9. Ibid., 22 March 1798.
10. *Manchester Mercury*, 16 April 1799.
11. 40 Geo III *c*. 39.
12. Huddersfield Canal Committee Minute Book, 11 September 1800.
13. Ibid., 18 December 1800.

14. Ibid., 13 December 1804.
15. 46 Geo III *c.* 12.
16. *Annual Register*, 1809, p. 343.
17. *Derby Mercury*, 18 April 1811.
18. Huddersfield Canal Committee Minute Book, 4 April 1811.
19. For this and other information we are indebted to articles by Mr Neil Fraser in the *Journal* of the Railway & Canal Historical Society in September 1961 and November 1962.
20. Peak Forest Canal Minute Book, 26 September 1811; Huddersfield Canal Committee Minute Book, 27 September 1811.
21. Huddersfield Canal Committee Minute Book, 22 October 1812.
22. Ibid., 24 June 1819.
23. Ibid., 25 April 1834.
24. Ibid., Proprietors' Minute Book, 24 June 1813.
25. *Report* of 1812.
26. *Report* of 1818.
27. *Report* of 1819.
28. *Report* of 1822.
29. Huddersfield Canal Committee Minute Book, 24 September 1817.
30. Ibid., Proprietors' Minute Book, 27 June 1833.
31. Ibid., Committee Minute Book, 4 February 1825.
32. Ibid., 30 January 1835.
33. Ibid., 23 June 1841.
34. Ibid., 28/29 June 1843.
35. Ibid., 19 September 1843.
36. See John Marshall, *The Lancashire & Yorkshire Railway*, Vol. I, 1969.
37. 8 & 9 Vic *c.* 105.
38. For early relations with the Sheffield, Ashton-under-Lyne & Manchester Railway and the Manchester, Sheffield & Lincolnshire Railway, see George Dow, *Great Central*, 1960, Vol. I, pp. 46, 49, 75–6, 78–9, 81, 95, 127.

Notes to Chapter XIII

1. JHC, 28 November 1721.
2. 8 Geo I *c.* 14.
3. For these schemes, see Charles Hadfield's forthcoming book, *The Canals of Yorkshire and the North East.*
4. William Chapman, *Second Part of a Report on the proposed Navigation between the East and West Seas*, 1795.
5. Report on meeting of 11 August 1818 (Bewley papers, Cumberland C.R.O. DX/182/3).
6. *Mr Chapman's Report on the means of obtaining a Safe and Commodious Communication from Carlisle to the Sea*, Carlisle, 1807 (I.C.E. Library).
7. William Chapman, *Report on the measures to be attended to in the Survey of a line of navigation from Newcastle upon Tyne to the Irish Channel*, 1795.
8. *Appendix to Mr Chapman's Report of June 27 1807* (I.C.E. Library).
9. *Mr Telford's Report on the intended Cumberland Canal: and Mr Chapman's further report on Observations thereon*, 1808 (I.C.E. Library).
10. *Mr Chapman's Report on the proposed Canal Navigation between Carlisle and Solway Frith*, 2nd ed., 1818 (I.C.E. Library).
11. 59 Geo III *c.* 13.
12. Carlisle Canal Minute Book, 27 December 1822.
13. *Address to the Subscribers to the Canal from Carlisle to Fisher's Cross*, 31 January 1823 (I.C.E. Library).
14. Carlisle Canal Minute Book, 28 October 1823.

15. *Observations on the most advisable measures to be adopted in forming a Communication . . . to or from Newcastle or Carlisle.*
16. For the Newcastle & Carlisle Railway, see K. Hoole, *A Regional History of the Railways of Great Britain, Vol. 4, The North East*, 1965, and John S. MacLean, *The Newcastle & Carlisle Railway*, 1948.
17. Carlisle Canal Minute Book, 1 July 1838.
18. For Houston's swift boats, see Jean Lindsay, *The Canals of Scotland*, 1968, pp. 92–3. For a description of a trip on the *Arrow*, see Chapter XXIII of Sir George Head, *A Home Tour through the Manufacturing Districts of England, in the summer of 1835*, 1836.
19. 6 & 7 Will IV *c.* 60.
20. Carlisle Canal Minute Book, 2 June 1836.
21. Ibid., 14 February 1845.
22. Jack Simmons, *The Maryport & Carlisle Railway*, 1947.
23. Carlisle Canal Minute Book, 9 April 1847.
24. Ibid., 6 July 1847.
25. Simmons, *Maryport & Carlisle Railway*, op. cit.
26. Carlisle Canal Minute Book, 6 March 1852.
27. Ibid., 15 April 1853.
28. The information on the Nent Force Level is entirely taken from Paul N. Wilson, 'The Nent Force Level', *Trans. Cumberland & Westmorland A. & A. Soc.*, Vol. LXIII, n.s., 1963, by kind permission of the author. For further material readers are referred to that paper.

Notes to Chapter XIV

1. Statement on the part of the Bridgewater Trustees, n.d. (*c.* 1850), Manchester Ship Canal (Bridgewater Dept) Records.
2. See John Marshall, *The Lancashire & Yorkshire Railway*, Vol. I, 1969, Chapter 7.
3. E. C. Mather, *After the Canal Duke*, Chapter X.
4. Statement on the part of the Bridgewater Trustees, op. cit.
5. Ibid.
6. *Manchester Guardian*, 11 October 1851.
7. For the material in this and the preceding paragraph we are indebted to Mr F. C. Mather's *After the Canal Duke*, Chapter XI and pp. 242–4.
8. 16 Vic *c.* 37.
9. See Charles Hadfield, *The Canals of the West Midlands*, 2nd ed., 1969.
10. 20 & 21 Vic *c.* 4.
11. Mather, *After the Canal Duke*, Chapter XIII.
12. Ibid.
13. Evidence on the Lancashire Union Railways Bill, 26 April 1864.
14. Mather, *After the Canal Duke*, Chapter VI.
15. *Haddock's Railway Time Table and Conveyance Directory*, December 1858.
16. *Powlson's Railway Time Table and Conveyance Directory*, September 1865.
17. *Report of E. Leader Williams, Jun . . . on the practicability of improving the Navigation of the River Weaver* (Manchester Ship Canal Museum).
18. Evidence before the Select Committee on Canals, 1883. Answer No 950.
19. Ibid., No 1071.
20. Ibid., No 1688.
21. *Warrington Guardian*, 16 July 1859.
22. *Railway Times*, 1 February 1873.
23. For the system, see Charles Hadfield, *The Canal Age*, pp. 188–9.
24. For the Manchester Ship Canal, see Sir Bosdin Leech, *History of the Manchester Ship Canal*, 1907, and Charles Hadfield, *British Canals*, 4th ed., 1969, Chapter XIV.
25. *Runcorn Guardian*, 3 June 1876.

26. *P.O. Directory of Cheshire*, 1878.
27. 38 & 39 Vic *c*. 91.
28. 26 Geo V & 1 Ed VIII *c*. 124.
29. 39 & 40 Vic *c*. 104.
30. See Manchester directories of 1874 to 1882.
31. Broadsheet, Manchester Central Library, M/CR f. 1882/3.
32. Minutes of Evidence on the Opposition by the Bridgewater Navigation Company and the Mersey and Irwell to the Manchester Ship Canal bill, 1885. House of Lords, Apps 24, 27, 28.
33. 48 & 49 Vic *c*. 188.

Notes to Chapter XV

1. See Charles Hadfield, *The Canals of the West Midlands*, 2nd ed., 1969.
2. Weaver Trustees' Minute Book, 5 June 1854.
3. Ibid., 2 July 1857.
4. Ibid.
5. Ibid., 26/27 June 1850.
6. Ibid., 24/25 June 1863.
7. *Report of E. Leader Williams Jun . . . on the practicability of improving the Navigation of the River Weaver so as to render it available for Sea-going Vessels*, Liverpool, 1865 (Manchester Ship Canal Museum).
8. Weaver Trustees' Minute Book, 6 June 1870.
9. See the account of the same period, from the North Staffordshire Railway's and Shropshire Union's point of view, in Charles Hadfield's *The Canals of the West Midlands*, 2nd ed., 1969, pp. 243–5.
10. Ibid., p. 174.
11. Ibid., pp. 140–2.
12. Charles Hadfield, *The Canals of the East Midlands*, 1970, 2nd ed., pp. 129–30.
13. Charles Hadfield, *The Canals of South West England*, 1967, pp. 93–4.
14. Ibid., pp. 104 et seq.
15. For a detailed description, see L. F. Vernon Harcourt, *Rivers and Canals*, 1896, II, 403–4, and *Procs. Inst. C.E.*, Vol. 45, 1876, p. 110.
16. See J. A. Saner, 'A Short Description of the River Weaver Navigation' in *Report of the Conference on Canals and Inland Navigation*, Royal Society of Arts, 1888.
17. Weaver Trustees' Minute Book, 24/25 July 1889.
18. For an account of the alterations, see 'Anderton Boat Lift', *The Engineer*, 24 July 1908.
19. See Charles Hadfield, *British Canals*, 4th ed., 1969, pp. 255 ff.
20. 27 & 28 Vic *c*. 296.
21. BTHR Gen 4/731 of 1928.
22. Information attributed to A. Ross of the Great Northern Railway, by J. H. Caine, 'Canals versus Railways', *The Great Central Railway Journal*, December 1908, p. 121.
23. See Barker, 'The Sankey Navigation', op. cit., p. 141.
24. Evidence of John Robinson on Bridgewater Canals Bill, 14 March 1906, House of Lords Committee.
25. Quo Barker, 'The Sankey Navigation', op. cit., p. 154.
26. Warrant of abandonment. Order of 20 October 1920.
27. Order of 21 May 1931.

Notes to Chapter XVI

1. Leeds & Liverpool Canal Committee Minute Book, 27 June 1848.
2. Minutes of lessees of Leeds & Liverpool Canal, 1850–7 (BTHR).
3. 27 & 28 Vic *c.* 288. See also Chapter XVII.
4. M. D. Greville, 'Chronological List of the Railways of Lancashire, 1828–1939', *Trans. Hist. Soc. Lancs & Ches*, Vol. 105, 1953.
5. Taken from 'Statistical History', op. cit. Blank years are due to gaps in the source tables.
6. Total revenue from all classes of goods carried, in addition to coal and merchandise.
7. For a full account, see M. D. Greville, 'West Lancashire Mysteries', Railway & Canal Historical Society *Journal*, Vol. 6, No. 3, May 1960; also John Marshall, *The Lancashire & Yorkshire Railway*, Vol. 1, 1969, p. 165.
8. Greville, op. cit.
9. M. D. Greville and G. O. Holt, 'Railway Development in Liverpool', *Railway Magazine*, February 1958.
10. 54 & 55 Vic *c.* 77.
11. Leeds & Liverpool Canal Board Minute Book, 17 August 1893.
12. Ibid., 15 March 1894.
13. Ibid., 17 January 1894.
14. Leeds & Liverpool Canal Minute Book, General Meeting, 24 March 1948.
15. *Second Report of the Commissioners for Inquiring into the State of Large Towns and Populous Districts*, Vol. II, 1845.
16. *Bradford Observer*, 1849–50.
17. Vint, Hill & Killick papers, Bradford Central Library.
18. Leeds & Liverpool Canal Minute Book, 13 January 1852–23 April 1852.
19. *Bradford Observer*, 1864–6.
20. Leeds & Liverpool Canal Minute Book, 16 January 1866.
21. *Bradford Observer*, 1866–7; Leeds & Liverpool Canal Minute Book, 16 January 1866–10 July 1867.
22. 34 & 35 Vic *c.* 155.
23. Prospectus, Bradford Canal Co Ltd, 1871 (G. J. Biddle collection).
24. Leeds & Liverpool Canal Minute Book, 5 September 1871.
25. *Bradford Observer*, 16 April 1873.
26. Mortgage deed, 30 April 1874, Vint, Hill & Killick papers, loc. cit.
27. 41 & 42 Vic *c.* 158.
28. Bradford Canal Joint Committee report, 31 December 1892, Vint, Hill & Killick papers, loc. cit.
29. Leeds & Liverpool Canal Minute Book, 15 May 1895.
30. Ibid., 18 September 1902.
31. Report to Joint Committee, 1898, Vint, Hill & Killick papers, loc. cit.
32. Evidence given during closure Bill proceedings, reported in *Yorkshire Observer* and *Canals & Waterways*, 1921–2.
33. T. R. Roberts, 'Bradford Waterway's Rise & Fall', *Bradford Textile Society Journal*, 1962–3.
34. From *Statement* in G. J. Biddle collection, 1888–92.
35. From Board of Trade returns.
36. From revenue accounts, 1903–22, Vint, Hill & Killick papers, loc. cit.
37. *Yorkshire Observer*, 27 April 1926.
38. 12 & 13 Geo V *c.* 29.

Notes to Chapter XVII

1. For the full history of the railway see M. D. Greville and G. O. Holt, *The Lancaster & Preston Junction Railway*, 1963, on which much of this account is based, supplemented by information from the Lancaster Canal Company's minute books (BTHR).
2. 6 & 7 Vic *c.* 4.
3. 7 & 8 Vic *c.* 37.
4. 12 & 13 Vic *c.* 87.
5. J. Barron, *History of the Ribble Navigation*, 1938.
6. These are fully detailed in M. M. Schofield, *Outlines of an Economic History of Lancaster, 1680 to 1860*, Part 2, 1951, and R. G. Armstrong, 'The Rise of Morecambe (1820–1862)', *Trans. Hist. Soc. of Lancashire and Cheshire*, Vol. 100, 1948, from which this account is taken.
7. Information from Mr M. M. Schofield.
8. Information from Mr M. D. Greville.
9. The committee approved a draft letter to Mr Cawkwell, General Manager of the L. & N.W.R.—Lancaster Canal Committee Minute Book, 26 September 1860.
10. 27 & 28 Vic *c.* 188.
11. 42 & 43 Vic *c.* 142.
12. 48 & 49 Vic *c.* 88.
13. For a full account see Paul N. Wilson, 'The Gunpowder Mills of Westmorland & Furness', *Trans. Newcomen Soc.*, XXXVI, 1963–4.
14. Schofield, *Economic History of Lancaster*, op. cit.
15. 1888 & 1898—Board of Trade returns; 1905—*Report of the Royal Commission . . . on Canals & Waterways*, 1909.
16. Information from Mr D. Ashcroft, who had a family carrying business on the canal which carried the last traffic.
17. 25 & 26 Geo V *c.* 49.
18. 2 & 3 Geo VI *c.* 28.
19. *Sunday Dispatch*, 23 January 1944.
20. Information from Mr D. Ashcroft.
21. According to J. D. Marshall, *Furness and the Industrial Revolution*, 1958. Number of vessels from *Jackson's Ulverston Almanac*, 1852.
22. H. F. Birkett, *The Story of Ulverston*, 1949.
23. *Jackson's Almanac*, op. cit.
24. W. White, *Furness Folk and Fact*, 1930.
25. 25 & 26 Vic *c.* 89.
26. Register of Arrivals & Departures, Ulverston Canal, January 1862–October 1879 (BTHR).
27. Canal Returns, 1888 and 1898; Canal Returns, Royal Commission on Canals and Waterways, 1905.
28. 8 & 9 Geo VI *c.* 2.

Notes to Chapter XVIII

1. Aire & Calder Navigation Directors' Minute Book, 6 October 1847.
2. Rochdale Canal Minute Book, 16 May 1849.
3. Rochdale Canal Report, 2 May 1850.
4. Rochdale Canal Minute Book, 10 October 1855.
5. Ibid., 11 January 1865.
6. Ibid., 10 November 1869.

7. Ibid., 20 January 1870.
8. Ibid., 13 May 1873.
9. Select Committee of the House of Lords on the Bridgewater Canals Bill, 14 March 1906.
10. 13 & 14 Geo V *c.* 77.
11. 15 & 16 Geo VI & 1 Eliz II *c.* 37.

Notes to Chapter XIX

1. We are indebted for information to Mr V. I. Tomlinson.
2. BTHR Gen 4/731. Also information from Mr John Marshall.
3. *Canal Returns*, 1888 and 1898. Royal Commission on Canals and Waterways, 1909, especially Vol. III, Appendix No. 10.
4. Report of the Royal Commission on Canals and Waterways, 1909, Vol. III, Appendix No. 10.
5. Ibid., Vol. I, Answer No. 4323.
6. Information from Alfred Gaskell's papers, *Th'Owd Cut* (1961) and *The Fletcher Canal* (1963) and from the author; and the Report of the Royal Commission, op. cit.
7. Report of the Royal Commission, op. cit., Answer No. 20,791.
8. *Canal Returns*, 1870.
9. Report of the Royal Commission, op. cit., Vol. I, Answer No. 1461.
10. Ibid., Answer No. 4324.
11. Ibid.
12. Winifred M. Bowman, *5000 Acres of Old Ashton*, 1950.
13. G. Dow, *Great Central*, 1959, Vol. I, pp. 158–9.
14. Ibid., Vol. I, pp. 42, 59.
15. See A. R. Bennett, *The Chronicles of Boulton's Siding*, 1927, pp. 15–16; *Lancashire & Cheshire Advertiser*, 3 May 1851; the *Hyde & Glossop News* for dates in 1858, and *A Guide to Marple, Lyme Park etc.*, 1845 (Stockport P.L.). We are grateful to Mr J. A. Hall for much of this information.
16. Joel Wainwright, *Reminiscences of a Life-time in Marple*, 1882, p. 14.
17. Brian Lamb, *The Bugsworth Complex* (MS), 1965.
18. Slater's *Manchester Directory*, 1848.
19. For much of the following material, in many cases taken from the *Huddersfield Examiner*, we are indebted to articles by Neil Fraser in the *Journal* of the Railway & Canal Historical Society for September 1961 and November 1962.
20. Minute 16,684 of London & North Western Railway Goods Conference, 12 June 1893.
21. Minute 16,868, ibid., 15 January 1894. We are indebted to Mr C. R. Clinker for these two references.
22. *Canal Returns*, 1870; Report of the Royal Commission, op. cit.

APPENDIX I

Summary of Facts about the Canals and Navigations of the North West

A. *Rivers Successfully Made Navigable*

River	*Date of Act under which Work was begun*	*Date Wholly Opened*	*Approx. Cost at Opening £*	*Terminal Points*
Douglas (alias Asland)	1720	1742	*c.* 7,900	Ribble estuary to Wigan
Mersey & Irwell	1721	1736	*c.* 18,000	Hunt's Bank, Manchester–Bank Quay, Warrington[2]
Weaver	1721	1732	16,000	Winsford–Frodsham bridge
		1765		Branch to Witton Mill
	1807	1810	50,000	Weston Canal (nr Sutton to Weston Point)
		c. 1783		Anderton branch

[1] Measured length on *A General Map of the Grand Canal from Liverpool to Leeds . . . from an accurate survey by Joseph Priestley*, n.d., Manchester Central Library. Drainage of coastal marshlands in the early *c.* 19 increased the length to about 19 m.

[2] Also Runcorn after the opening of the Runcorn & Latchford Canal in 1804.

[3] In its final state, from Hunt's Bank to Runcorn.

[4] Originally 8, later 9. Two more were added with the Runcorn & Latchford Canal in 1804, making total 11. Reduced to 10 in 1883 when Calamanco lock was eliminated. These figures include Howley lock, which was not on the main line after the Runcorn & Latchford Canal was opened.

B. *Rivers with Uncompleted Navigation Works*

None.

Length	*Greatest Number of Locks*	*Size of Boats Taken*	*Date of Disuse for Commercial Traffic*	*Date of Abandonment*	*Whether bought by Railway and Present Ownership*
17½[11] miles	8		1780–1801		No. Bought by Leeds & Liverpool Canal Company
28¾ miles[3]	11[4]	68 ft × 17 ft 6 in.	c. 1887[5]	1885[5]	Bought by Lord Francis Egerton, 1844; transferred to Bridgewater trustees, 1846. Manchester Ship Canal Co
18⅜ miles[6]	11[7]	[10]			British Waterways Board
¾ mile	1[8]	68 ft × 16 ft 9 in.[11]			
4 miles ⅛ mile	3[9]				

[5] As a whole. Portions not built over for the Manchester Ship Canal remained in use.
[6] Shortened later at various times to 17 miles (16 miles to the junction of the Weston Canal, and 1 mile thence to Frodsham bridge).
[7] Reduced over the years to the present 4.
[8] Eliminated 1829.
[9] Now 2, one of which is normally level.
[10] The size of craft taken when the navigation was first built is not known, except that locks were between 15½ and 16 ft wide.
[11] After the rebuilding of the 1760s.

C. *Canals, the Main Lines of which were Completed as Authorized*

Canal	*Date of Act under which Work was begun*	*Date wholly Opened*	*Approx. Cost at Opening £*	*Terminal Points*	*Branches Built*
Ashton-under-Lyne	1792	c. 1796[1]	170,000	Piccadilly, Manchester–Ashton	
	1793	1797			Stockport Beat Bank
		1796			Hollinwood
		1797			Fairbottom
		c. 1796			Werneth[5]
					Islington
					Dukinfield
Bradford	1771	1774		L. & L. Canal, Shipley–Bradford	None
Bridgewater	1759	1765	c. 7,500	Worsley–Salford and Worsley–Hollins Ferry[10]	
	1760	1761			Stretford–Longford bridge
	1762	1776	c. 70,000		Longford bridge–the Hempstones[13]
		c. 1799			Leigh
		c. 1773			Preston Brook
		1838			Hulme locks
		before 1785			Moss
	1853	1859	c. 40,000		Runcorn & Weston

[1] Except for the Ancoats–Piccadilly section, completed in 1799.
[2] Unfinished.
[3] Including a staircase pair at Waterhouses.
[4] Part.
[5] A private canal from the end of the Hollinwood branch to Old Lane colliery, Oldham.
[6] 1774–1867.
[7] & [8] The effect of a High Court injunction.
[9] 1872–1922. The authorizing Act was 1871, closing the last $\frac{3}{8}$ mile in Bradford, below which the canal was reopened to Oliver lock in 1872 and to Northbrook bridge in 1873.

Length	*Greatest Number of Locks*	*Size of Boats Taken*	*Date of Disuse for Commercial Traffic*	*Date of Abandonment*	*Whether bought by Railway and Present Ownership*
6¾ miles	18	70 ft × 7 ft	1957		Bought by Manchester, Sheffield & Lincolnshire Railway, 1848
4⅞ miles			1933	1962	
3 miles[2]				1798	British Waterways Board
4⅝ miles	7[3]		1932	1955[4]	
				1961	
1⅛ miles			1930	1955	
1 mile					
¼ mile					
c. 40 yd					
3⅜ miles[6]	10	61 ft × 14 ft 4 in.	1867[7]	1867[8]	No. Bought jointly by Leeds & Liverpool and Aire & Calder
3 miles[9]			1922	1922	
7¼ miles[11]	None	72 ft × 14 ft 2 in.[12]	Open		Manchester Ship Canal Co.
½ mile			Open		
	10[14]		Open[15]	15	
25¼ miles					
6¼ miles			Open		
¾ mile			Open		
⅛ mile	3[16]		Open		
c. ⅝ mile					
1¼ miles		72 ft 3 in. × 18 ft 5 in.	1962	1966	

[10] Altered 1760 to Worsley–Manchester on an amended line. About 2 miles of the Hollins Ferry line was built, from which a ⅞ mile branch was cut *c.* 1800.

[11] The Worsley underground canals are not included here. They extended for some 46 miles, and finally closed in 1887.

[12] The size of boat used in the underground canals and also on the main canal was 47 ft × 4½ ft.

[13] The Hempstones changed by the Trent & Mersey Act of 1766 to Runcorn.

[14] In three staircase pairs. A second flight was opened *c.* 1827.

[15] Except for old Runcorn locks, abandoned 1949, and new Runcorn locks, abandoned 1966.

[16] Converted to a single lock, 1962.

Canal	*Date of Act under which Work was begun*	*Date wholly Opened*	*Approx. Cost at Opening £*	*Termina Points*	*Branches Built*
Carlisle	1819	1823	81,000	Port Carlisle–Carlisle	
Fletcher's	None	*c.* 1791		M.B. & B. Canal at Clifton–Fletcher's collieries	
Huddersfield	1794	1811	400,000	Ashton Canal–Sir John Ramsden's C. (Huddersfield)	
Leeds & Liverpool	1770	1816	824,000[4]	R. Aire, Leeds–Liverpool	
		1781			Rufford (Lower Douglas)
		c. 1862[9]			Rain Hall Rock
		c. 1800			Tunnel (Crooke) (private)
					Ince Hall Coal and Cannel (private)
		1820			Leigh
		1846			Stanley Dock
		1845			Arches Lock (Monk Pits)

[1] Short narrow boats able to work on the Manchester, Bolton & Bury Canal in pairs, were used on this canal.

[2] Except for short section at Huddersfield transferred to Calder & Hebble.

[3] The section referred to in note 2.

[4] Up to 1844, including branches but excluding improvements, Arches lock and Stanley dock branches.

Length	*Greatest Number of Locks*	*Size of Boats Taken*	*Date of Disuse for Commercial Traffic*	*Date of Abandonment*	*Whether bought by Railway and Present Ownership*
$11\frac{1}{4}$ miles	8	74 ft × 17 ft 6 in.	1853	1853	Company turned itself into a railway company 1853
$1\frac{1}{2}$ miles	1	68 ft × 7 ft[1]	1935		
$19\frac{7}{8}$ miles	74	70 ft × 7 ft	1944	1944[2] 1963[3]	Sold to Huddersfield & Manchester Railway in 1844 British Waterways Board
$127\frac{1}{4}$ miles[5]	92[6]	72 ft × 14 ft[7] 62 ft × 14 ft[8]	Open	Open	No. British Waterways Board
$7\frac{1}{4}$ miles	8	72 ft × 14 ft	Open	Open	
$\frac{3}{8}$ mile		62 ft × 14 ft	1891		
$\frac{1}{4}$ mile		72 ft × 14 ft	c. 1940		
$\frac{1}{2}$ mile		62 ft × 14 ft	c. 1950		
$7\frac{1}{4}$ miles	4	72 ft × 14 ft	Open	Open	
$\frac{1}{4}$ mile	4	72 ft × 14 ft	Open	Open	
90 yd	1	62 ft × 14 ft	c. 1960		

[5] Including $10\frac{3}{4}$ miles of the South End of the Lancaster.
[6] Including later duplications and additions.
[7] Liverpool–Wigan–Leigh and Stanley Dock branch.
[8] Wigan–Leeds.
[9] Commenced 1796.

Canal	*Date of Act under which Work was begun*	*Date wholly Opened*	*Approx. Cost at Opening £*	*Terminal Points*	*Branches Built*
Manchester, Bolton & Bury	1791	1808	115,500	R. Irwell, Manchester–Bolton	
					Bury
Manchester & Salford Junction	1836	1839	*c.* 60,000	R. Irwell at Water St–Rochdale Canal	
Manchester Ship	1885	1894	14,350,000	Eastham–Salford	
Ravenhead	None	*c.* 1773		Thatto Heath–Ravenhead, St Helens	
Rochdale	1794	1804	600,000	Bridgewater C. (Castlefield)–Calder & Hebble Nav. (Sowerby Bridge)	
		1798			Rochdale
		1834			Heywood

[1] The 6 from the Irwell to Oldfield Road included 2 staircase pairs; the flight of 6 at Prestolee was in 2 staircases of 3 locks each.
[2] To Bolton.
[3] $6\frac{7}{8}$ miles abandoned.
[4] Remainder of canal.
[5] To Bury.
[6] Disuse of the canal except for transit at Bury wharf, which ended in 1956.
[7] $\frac{3}{8}$ mile after partial abandonment in 1875.

Length	*Greatest Number of Locks*	*Size of Boats Taken*	*Date of Disuse for Commercial Traffic*	*Date of Abandonment*	*Whether bought by Railway and Present Ownership*
11 miles 4¾ miles	17[1]	68 ft × 14 ft 2 in.	Before 1924[2] 1936[5] 1950[6]	1941[3] 1961[4]	Turned themselves into a railway company also, 1831. Amalgamation with Manchester & Leeds Railway, 1846. British Waterways Board
⅝ mile[7]	4[8]	71 ft × 14 ft	*c.* 1922	1875[9] 1936	Bought by Mersey & Irwell, 1842
36 miles	5[10]	600 ft × × 65 ft[11]			Manchester Ship Canal Co.
c. ½ mile	None				
33 miles ½ mile 1½ miles	92	74 ft × 14 ft 2 in.	1937[12] 1958[14]	1952[13]	Rochdale Canal Co.

[8] The upper three were duplicated, and the upper two were staircase pairs.
[9] Between Lower Mosley Street and Watson Street.
[10] Four pairs of locks, with three at Eastham. There are also three side locks into the Mersey.
[11] Size of locks.
[12] The last boat over the full length of the canal.
[13] Except a section in Manchester (see text).
[14] The last working boat passed on 29 May 1958 from Bloom Street, Manchester, power station, to the Bridgewater Canal.

Canal	*Date of Act under which Work was begun*	*Date wholly Opened*	*Approx. Cost at Opening £*	*Terminal Points*	*Branches Built*
Sankey Brook (St Helens)	1755	1757	18,600	Sankey Bridges–Old Double Lock	
		1762			Blackbrook[4]
		1759			Gerard's Bridge
		c. 1772			Boardman's Bridge and Ravenhead
	1762			Sankey Bridges–Fiddler's Ferry	
	1830	1833	30,000	Fiddler's Ferry–Widnes	
Springs Branch (Lord Thanet's)	1773	1797		L. & L.C., Skipton–Haw Bank Quarry Staithes	None
Ulverston	1793	1796	Over 9,200	R. Leven estuary–Ulverston	None

[1] Including one staircase pair. The river lock at Sankey Bridge was probably disused soon after 1830.

[2] To St Helens.

[3] North of Newton Common lock.

Length	*Greatest Number of Locks*	*Size of Boats Taken*	*Date of Disuse for Commercial Traffic*	*Date of Abandonment*	*Whether bought by Railway and Present Ownership*
8 miles	10[1]	68 ft × 16 ft 9 in.	1919[2] 1959	1931[3] 1963	Amalgamated with St Helens & Runcorn Gap Rly Co. in 1845. British Waterways Board
⅝ mile				1931	
1½ miles				1931	
1⅝ miles	2[5]			1898[6] 1920[6] 1931	
1⅝ miles	1			1963	
3⅜ miles	1[7]	75 ft × 20 ft		1963	
½ mile	None	62 ft × 14 ft	Open	Open	
1½ miles	1	104 ft × 27 ft	1917	1945	Furness Rly, 1862. Ulverston U.D.C.

[4] There was a small private canal about ½ mile long between Blackbrook and Carr Pool. See text.

[5] A staircase pair.

[6] Parts of Ravenhead arm.

[7] Two locks side by side into the river.

D. *Canals, the Main Lines of which were not Completed*

Canal	*Date of Act under which Work was begun*	*Date Opened*	*Approx. Cost at Opening £*	*Authorized Terminal Points*	*Terminal Points as Built*
Lancaster	1792	1819		Kendal–Westhoughton	Kendal–Wigan (Top Lock, L. & L. Canal)
	1792	1825			
	1819	1816			
Peak Forest	1794	1800[4]	117,000[5]	Ashton Canal (Dukinfield)–Chapel Milton[6]	Ashton Canal (Dukinfield)–Bugsworth[7]

[1] Excluding intervening tramroad, Preston–Walton Summit, 5 miles. This is made up of 57 miles of the North End (Kendal–Preston), 3 miles of the Walton Summit line, and 10¼ miles from Johnson's Hillock to Wigan top lock.
[2] Transport Act 1955 authorized closure to navigation, but section Preston (Ashton basin) to Tewitfield bottom lock including Glasson branch still open to pleasure craft.
[3] Formed part of Leeds & Liverpool main line and leased to them with South End of Lancaster, 1863.

E. *Canals Partly Built but not Opened*

Ashton-under-Lyne Canal. The Beat Bank Branch off the company's Stockport line, 3 miles long, was authorized in 1793, partially built, and abandoned in 1798.

F. *Canals Authorized but not Begun*

1721 *Dane River.*
1766 *Bridgewater Canal*, Stockport branch.
1794 *Haslingden Canal*, from the Manchester, Bolton & Bury Canal at Bury to Accrington (Church).

Branches Built	*Length*	*Greatest Number of Locks*	*Size of Boats Taken*	*Date of Disuse for Commercial Traffic*	*Date of Aban-donment*	*Whether bought by Railway and Present Ownership*
	67¼ miles[1]	8	72 ft × 14 ft 6 in.	1947	1955[2]	L.N.W.R. 1885. British Waterways Board
Glasson Dock	2½ miles	6	72 ft × 14 ft 6 in.	1947	1955[2]	
Johnsons Hillock[3]	½ mile	6	62 ft × 14 ft	Open	Open	
	14¾ miles	16	70 ft × 7 ft	1959		Leased to the Sheffield, Ashton-under-Lyne & Manchester Rly in 1846
Whaley Bridge	⅜ mile			1926[8]		British Waterways Board

[4] But with a connecting tramroad at Marple. The locks were opened in 1804.

[5] To the opening in 1800 with a connecting tramroad at Marple. The locks cost another £27,000.

[6] The line from Bugsworth to Chapel Milton was built as part of the Peak Forest tramroad.

[7] Now Buxworth.

[8] From Buxworth to junction with Whaley Bridge branch.

APPENDIX II

Principal Engineering Works

A. *Inclined Planes*

Canal	*Name*	*Vertical Rise*	*Dates Working*	*Notes*
Bridgewater (Worsley underground canals)		*c.* 105 ft	1797–1822	Double-track counter-balanced, mine boats being carried dry in cradles. Twin locks at the head

B. *Lifts*

Canal	*Name*	*Vertical Rise*	*Dates Working*	*Notes*
Weaver	Anderton	50 ft	1875–present	Originally water-counterbalanced, assisted by steam power. Changed to electric power, 1903. Converted to independent operation of caissons using counter-balance weights, 1908

C. *Tunnels over 500 yards*

Huddersfield Canal	Standedge	5,456 yd (now 5,698 yd)
Leeds & Liverpool Canal	Foulridge	1,640 yd
Leeds & Liverpool Canal	Gannow	559 yd

D. *Outstanding Aqueducts*

Ashton-under-Lyne Canal	Ancoats (road)
Ashton-under-Lyne Canal	Dukinfield (Tame)
Ashton-under-Lyne Canal	Waterhouses (Medlock)
Bridgewater Canal	Barton Swing (M.S.C.)
Lancaster Canal	Brock
Lancaster Canal	Lune (Lancaster)
Lancaster Canal	Keer (Capernwray)
Lancaster Canal	Wyre (Garstang)
Leeds & Liverpool Canal	Dowley Gap (Aire)
Leeds & Liverpool Canal	Priestholme (Aire)
Manchester, Bolton & Bury Canal	Clifton (Irwell)
Manchester, Bolton & Bury Canal	Damside (Tonge)
Manchester, Bolton & Bury Canal	Prestolee (Irwell)
Peak Forest Canal	Marple (Goyt)
Rochdale Canal	Hebden Bridge (Calder)

INDEX TO VOLUMES I AND II

The principal references to canals and river navigations are indicated in bold type